INTRODUCTION TO ORE DEPOSITS

INTRODUCTION TO ORE DEPOSITS

LUDWIG BAUMANN

Professor of Economic Geology
Bergakademie Freiberg
East Germany (D.D.R.)

A HALSTED PRESS BOOK

JOHN WILEY & SONS
NEW YORK

PUBLISHED BY
SCOTTISH ACADEMIC PRESS LTD.
25 PERTH STREET, EDINBURGH EH3 5DW

First published 1976

Published in the U.S.A. by
HALSTED PRESS, a Division of
John Wiley & Sons, Inc., New York.

© Ludwig Baumann 1976

All rights reserved. No part of this publication may be reproduced, stored in a retrieval system, or transmitted, in any form, or by any means, electronic, mechanical, photocopying, recording or otherwise, without the prior permission of the Scottish Academic Press Ltd., 25 Perth Street, Edinburgh EH3 5DW.

Library of Congress Cataloging in Publication Data

Baumann, Ludwig.
 Introduction to Ore Deposits.

 "A Halsted Press book."
 Bibliography: p.
 1. Ore-deposits. 2. Geology, Economic. I. Title.
 TN263.B29 1975   553'.1   74–26527
 ISBN 0–470–05937–0

Printed in Great Britain by
R. and R. Clark Ltd., Edinburgh

Preface

The idea of publishing *Introduction to Ore Deposits* as a book was suggested by the interest expressed during my stay, in 1969, as visiting professor at the University of St. Andrews in many talks with earth scientists, and students on questions of economic geology. After the sudden and untimely death of Professor C. F. Davidson I was asked to deliver a course of lectures on 'The Science of Mineral Deposits' at the Department of Geology of the time-honoured University of St. Andrews. In a course of ten lectures an introduction was given to the process of deposit formation with special regard to the formation of ore deposits. The present book is based upon these lectures.

It is not my intention to write a new book on the science of mineral deposits, but with this *Introduction to Ore Deposits* I want to convey the basic knowledge of this branch of science to students of geology and economic geology, of mineralogy, geochemistry, petrology as well as of geography and mining.

Besides general information on the science of mineral deposits, the most important geological and physico-chemical conditions of formation of the genetic types of deposits of the endogenetic and exogenetic cycles as well as metamorphic transformation are dealt with. In writing the book it was assumed the reader was already acquainted with the fundamentals of geology, mineralogy and geochemistry. Therefore, the basic terms and definitions of these special fields are not repeated here. The general fundamentals of the science of mineral deposits are treated first and some examples given of typical ore deposits. Some well-known deposits from all over the world are described together with less familiar deposits from Central Europe.

For encouragement and steady support during the preparation of this book for printing I am deeply grateful to the Head of the Department of Geology of St. Andrews, Professor E. K. Walton. For the translation and steady assistance in proof reading of the typed copy I express my cordial thanks to the Head of the Language Department of the Bergakademie Freiberg, Mr H. Petzschner. I further record my sincere thanks to Dr P. Bowden, St. Andrews, for his valuable advice and assistance in the preparation of the manuscript as well as to the Scottish Academic Press.

LUDWIG BAUMANN

Contents

Introduction   1

1. Formation of ore deposits by magmatic processes (endogenetic cycle)   6
    1.1. The magma and its origin   6
    1.2. Magmatic differentiation   8
    1.3. Products of the differentiation of the magma   10
       1.3.1. Liquid-magmatic (intramagmatic) formation of deposits   13
       1.3.2. Pegmatitic-pneumatolytic formation of deposits   20
       1.3.3. Hydrothermal formation of deposits   28
    1.4. Spatial distribution of magmatic deposits   48
       1.4.1. Zonal distribution   48
       1.4.2. Level of intrusion   50
       1.4.3. Area of precipitation   51
       1.4.4. Tectonics (regional, local)   52
    1.5. Ore deposits, tectonics and magmatic activity   57
       1.5.1. The geotectonic-geomagmatic cycle   57
       1.5.2. The types of ore deposits of simatic magmatism   60
       1.5.3. The types of ore deposits caused by sialic magmatism   61

2. Formation of ore deposits by sedimentary processes (exogenetic cycle)   63
    2.1. The importance of the hydrosphere, atmosphere and biosphere   63
    2.2. Fractionation and deposition   64
       2.2.1. Weathering and fractionation of the elements   64
       2.2.2. Transportation and deposition   68
    2.3. Formation of ore deposits of the exogenetic cycle   75
       2.3.1. Ore deposits formed by weathering   75
       2.3.2. Precipitated sedimentary ore deposits   86

3. **Metamorphic transformation of ore deposits** — 104
   3.1. Causes and factors acting in metamorphism — 104
   3.2. Kinds of metamorphism and its products — 105
   3.3. The formation of metamorphic ore deposits (regenerated and mobilized deposits) — 106
      3.3.1. Transformation by static-kinetic metamorphism — 107
      3.3.2. Transformation by contact metamorphism — 111
      3.3.3. Polymetamorphic ore deposits — 112

Bibliography — 114

Appendix
   Metal-content of the most important ore minerals — 117

Index — 127

Introduction

Economically important deposits can be treated in different ways by a number of branches of science and technology including geology, mineralogy, mining, ore-dressing, metallurgy, economics and planning. For all branches of mining the deposit is the 'origin of all things', and effort is directed to the discovery and mining of deposits as a prerequisite of their utilization. The geologist aims at understanding the origin of ore deposits, partly as an aid to discovering further deposits, partly as a set of clues in general earth-history. This text deals briefly with the principles of ore-deposit formation. These, it is hoped, emerge by abstracting, from the wealth of material available, the most important preconditions governing processes involved in and relations amongst ore-deposit formation. For illustration a few of the world's best-known or most important deposits are described.

Definition

By the word 'deposit' we mean here a natural accumulation of useful materials in the earth's crust that can be won with economic profit. The last factor introduces a dependence on technological and social conditions. Such accumulations do not only include metals but also other raw products that are useful because of their chemical and physical properties. According to their main useful constituent we call the deposits: ore deposits, coal deposits, petroleum deposits, salt deposits or non-metallic deposits.

Ore deposits consist of definite *'mineral aggregations'* or *'mineral associations'* or *'mineral parageneses'*. Most mineral deposits are of relatively small size as compared with the whole of the earth's crust. But there are also mineral associations, consisting mainly of silicates, that extend over wide areas and play a principal role in the structure of the solid earth's crust. In these cases the 'mineral deposit' is the same as the 'rock', and the study concerned mainly with the rocks is known as petrography.

On the other hand, ore deposits and rocks are, besides their mineral composition, *'geological bodies'*. Their shape, kind and composition as well as, in part, their origin are inseparably connected with the geological development of the region in which they occur and it is only in this setting that they can be studied and understood.

The science of deposits—economic geology—is a special discipline standing between mineralogy and geology and using the methods of both sciences. Mineralogical methods play a decisive part in determining the composition of a deposit, its workability, and its properties during processing as well as giving some indication of its genesis. But the investigation of the place of formation, the size and shape of a deposit, which is first of all important for the methods of exploration and mining, is mainly a geological task (Fig. 1).

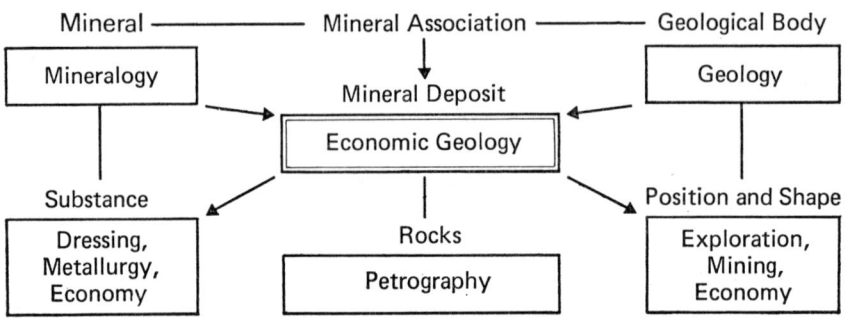

FIG. 1. 'Economic Geology' and its relations to some other mining sciences.

Ore deposits consist of mineral associations from which certain metals or metallic compounds can be extracted. Thus an 'ore' is always a mixture of mineral grains, either of the same kind or, more usually, of different species. The metal-bearing minerals are called 'ore minerals' whereas their nonmetallic accompanying minerals are called 'gangue' (for instance, galena +quartz or carbonates ='ore'; chromite +serpentine = 'ore'; gold +quartz ='ore'). 'Country rock' surrounds the deposit, contains little or no ore and is composed of rock-forming minerals.

Workability

The term 'ore' must be closely connected with the term 'workability'. The workability of an ore is very different according to the value of the ore and depends on many factors. The two chief factors are:

(*a*) the degree of concentration, that is the *metal content* (tenor) of a deposit, and
(*b*) the *size* of a deposit.

Moreover, the workability of a deposit depends on geographic factors (location, traffic conditions), technological factors (degree of industrialization) and finally politico-economic factors (demand of the national economy). Of course, all these factors may have a great influence on each other. Thus, for instance, the limit of workability of an ore in countries where engineering is less developed will lie at a higher metal content as compared with those having a highly developed technology. Here, attention should be paid to the fact that where the lower limit of the metal content is always high, the cheaper is the metal in question. Therefore, for various metals to be workable a certain minimum concentration is necessary (Fig. 2). The average content in the uppermost part (15-20 km) of the earth's crust of the individual chemical elements is the abundance ratio or the Clarke number. The technical usability of metallic raw materials has as a precondition strong local concentrations of the extremely small average amounts (mostly from 100- to 1000-fold).

The size of a deposit also has a decisive influence on the lower limit of workability. The larger the metal reserves of a deposit, the more money can be invested for technical installations. Unit-costs will thereby be lowered and, consequently, also the lower limit of workability.

Scope of economic geology

The *scientific* task of the investigation of ore deposits is to examine and explain the composition, shape and genesis of metal concentrations within the earth's crust. Its *practical* purposes are the economic extraction and optimal utilization of the content of the deposit (this comprises mining of raw materials and the treatment of the raw materials by ore dressing and metallurgical processes). A further aim is the recognition of criteria which allow the prediction and proving of further accumulations of useful minerals within the earth's crust

Metal	Average in earth's crust		Minimum content in workable deposits 1)	Factor of concentration
	g/t	%		
Al	813000	8,13	30 %	3,7 x
Fe	50000	5,00	25–30 %	5–6 x
Mn	1000	0,1	35 %	350 x
Cr	200	0,02	30 %	1500 x
Ni	80	0,008	1,5 %	188 x
Zn	80	0,008	4 %	500 x
Cu	70	0,007	1 %	140 x
Sn	40	0,004	1 %	250 x
Pb	16	0,0016	4 %	2500 x
Ag	0,1	0,00001	500 g/t	5000 x
Au	0,005	0,0000005	5–10 g/t	1000–2000 x
	1) Deposits in solid rock, no placers			

FIG. 2. Average contents of some important metals of the earth's crust and their minimum amounts to be workable deposits (after Schneiderhöhn, 1962).

accessible to man (this also involves methods of exploration and the opening-up of deposits).

Thus, the exact investigation of an ore deposit is a many-sided task, which requires a thorough knowledge of mineralogy, microscopy, petrography, chemistry, geology and tectonics (structural geology). The deposit has to be considered from the most varied viewpoints. In so doing it should always be kept in mind that a deposit within the earth's crust is not 'a thing in itself', but a member in a series of phenomena and causes, which have to be thoroughly studied and recognized.

Classification

Different principles may be used for classifying deposits.

(*a*) Classification based on *shape* (structure): this is especially advantageous for the practice of mining and for exploration methods. According to this division we can recognize *sheetlike* deposits, which are primarily characterized by their two-dimensional extension (for instance conformable, syngenetic seams; lens-shaped deposits; epigenetic lodes and veins). *Stocks* are mostly large, three-dimensional ore bodies with structures unconformable to the enclosing rock and more or less sharp boundaries (greisen bodies, metasomatic replacement bodies). *Impregnations* or disseminated ores (nests, 'schlieren') are porespace fillings in unaltered or only partly altered country rock. Such zones of impregnation are closely related to stocks and lens-shaped deposits. They show, however, gradual transitions to the unchanged original rock.

(*b*) Classification based on the *useful metal*. This division is particularly useful for the metallurgist, ore-dresser and economist. The main divisions with their common elements are:
— Iron and ferro-alloy metals (Fe, Mn, Cr, Ni, Co, Mo, W, V, Ti);
— Precious metals (Au, Ag, Pt and Pt companions);
— Nonferrous metals (Cu, Zn, Pb, Sn);
— Light metals (Al, Mg, Be);
— Special metals (Hg, Sb, As, Bi, U, rare earths etc.).

(*c*) Classification on a *regional basis* (particularly orogenic zones and metallogenetic provinces): the epochs and zones of large-scale mountain building are simultaneously predestined for abundant formation of deposits ('To the orogen belongs the chalcogen'); for example, the ore-deposit provinces of the Caledonian orogenesis, the Variscan orogenesis, the Alpine orogenesis. In Europe the formation of the chief deposits moves from north to south with the orogenic zones. To each orogenic belt belongs a 'metallogenetic province', which may be further divided into sub-provinces or ore-districts.

(*d*) Classification based on *genesis*. This classification is based upon 'isogenetic mineral paragenesis', that is the mineral associations that originated about the same time and approximately under the same physicochemical conditions in the same place. These isogenetic parageneses and the deposit-forming processes upon which they are based can be divided strictly genetically. According to the geologic events occurring on the earth, the formation of ore deposits takes place in two great cycles, namely an interior (endogenetic) cycle and an exterior (exogenetic) cycle. In both cycles significant fixations and displacements of elements occur. We distinguish between a magmatic process, a sedimentary process and a metamorphic process. Deposit formation is bound-up with each of these processes.

The origin of *magmatic deposits* is connected with a magma, its solidification and the processes associated with it.

The *sedimentary deposits* form under the influence of the hydrosphere, biosphere and atmosphere during weathering and under definite conditions of sedimentation.

The so-called *metamorphic deposits* were originally either magmatic or sedimentary deposits, which were transformed by changed physical and chemical conditions.

1
Formation of ore deposits by magmatic processes (endogenetic cycle)

1.1. The magma and its origin

Structure of the earth's crust: the first processes of separation of the elements must have taken place during the gaseous-liquid states of the earth. As regards today's structure of the earth, many physical facts testify to a strong physical and chemical or mineralogical inhomogeneity of the earth in the form of a world-wide division into shells. The specific gravity of the surface rocks is about 2·8, that of the whole earth about 5·5. Thus, in the earth's interior the surface rocks must be replaced by considerably denser material. Seismology allows inferences to be made about the inaccessible layers of the earth by means of the velocity of earthquake waves, which reveal sudden changes at various depths. They confirm that we may assume an upper layer, the Upper Crust or Sial, in which the elements Si and Al predominate (density about 2·8). This is underlain by other layers, the Lower Crust and the Upper Mantle (Sima), in which Fe and Mg are important with Si (density about 3·8). There are distinctly marked separating surfaces or discontinuities. One of them is the *Conrad* discontinuity, which separates the Upper from the Lower Crust. Within this zone, reaching a depth of 20-25 km, considerable melting of the mostly granite-like (acid) Upper Crustal material may take place at temperatures of 600-700 °C, leaving a basic residue of dioritic-gabbroid composition.

The *Mohorovičič* discontinuity is located at a depth of about 40 km. It marks a very distinct change in density and represents the boundary region between the feldspar-bearing rocks (gabbro) and the pyroxenites and peridotites not containing feldspar. It is below this boundary, that is in the Mantle, that basaltic magmas of juvenile origin may differentiate, at a temperature of about 900 °C.

Finally, the peridotite shell is underlain by the Griquaite or Gutenberg zone at a depth of about 60-100 km, with the *Mintrop* discontinuity between them. This zone, which is derived from crystallization and differentiation of the original crust of the earth, is normally a crystallized high-pressure facies, which changes to basaltic magma only when the pressure is released (especially during the early geosynclinal stage in the form of deep-reaching zones of tension). The rising basaltic

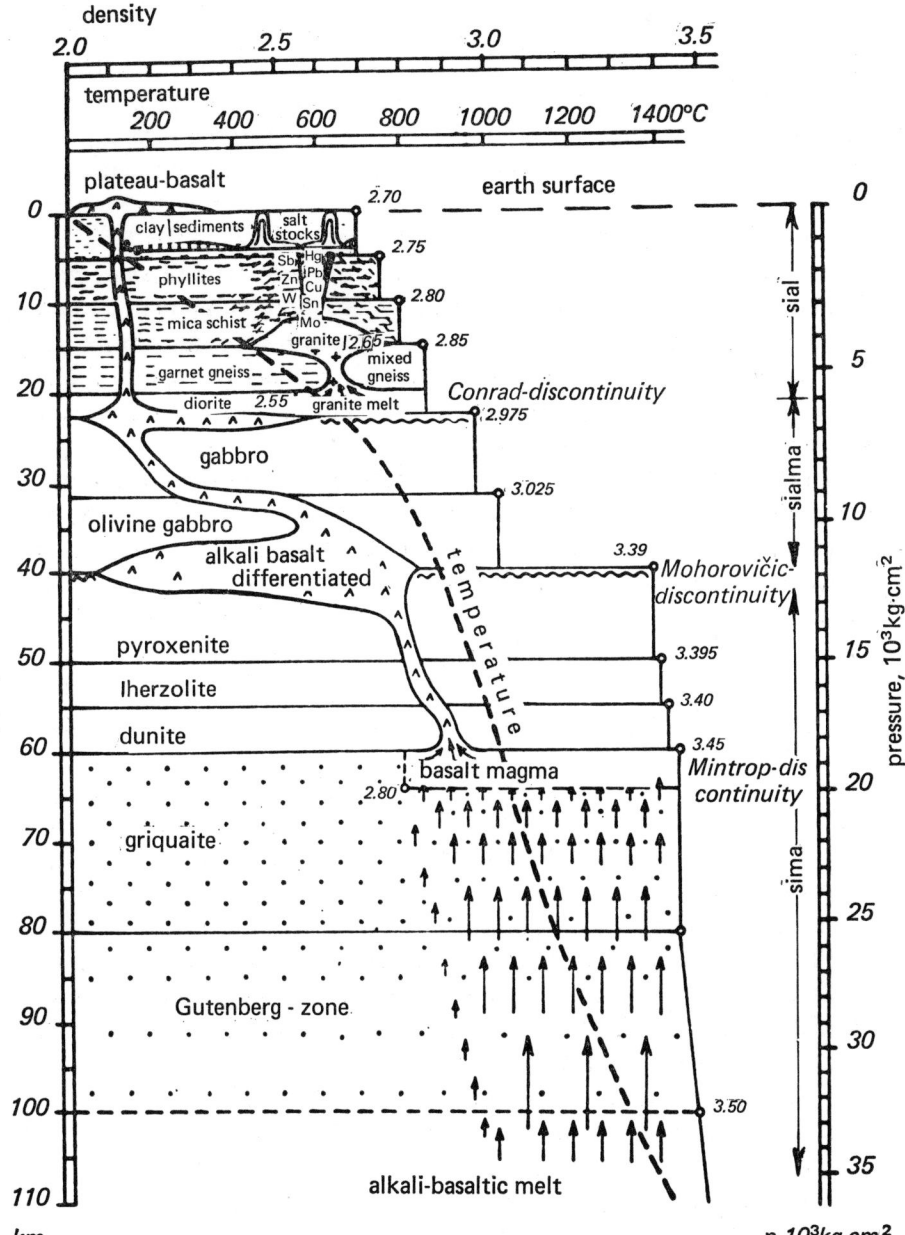

FIG. 3. Division of the earth's crust according to petrological and geophysical viewpoints (after Borchert 1960).

magma then forms larger reservoirs of magma, chiefly along the changes in density of the Mohorovičić and Conrad discontinuities. In these intermediate reservoirs, upon slow cooling, fractional crystallization takes place yielding more basic and acid components (Fig. 3).

Kinds of magma. Accordingly, there are two main types of magma. The acid sialic magmas occur in the range of the Conrad discontinuity, mostly only in connection with orogenesis. These granitic magmas are assumed to have originated by melting of part of the crust due to sinking down to great depths in geosynclines and by the frictional heat during folding. Therefore, these magmas are also termed *sialic-palingenetic* magmas. Due to the lower density of the granitic melt (2·55) as compared with the overlying rocks (gneiss =2·85) an uprise and intrusion of the magma takes place. This process represents the so-called synorogenic (plutonic) and subsequent magmatism (after Stille, 1940).

In contrast to these sialic magmas of the orogeny the basic simatic magmas occur chiefly in the early, geosynclinal stage. Here they are mainly, in the zones of greatest subsidence, bound up with deep-reaching zones of tension. The resulting pressure-relief brings about liquefaction, an increase in volume, and a decrease in density. Thus, uprising and differentiation of the magma in the range of the Mohorovičić and Conrad discontinuities become possible. The uprise of these *simatic-juvenile* magmas has been called initial magmatism by Stille.

1.2 Magmatic differentiation

When the magma rises to higher zones within the earth's crust, a separation of material takes place during crystallization, which is generally called magmatic 'differentiation'. This differentiation results in the formation of various igneous rocks, on the one hand, and different mineral and ore deposits, on the other. The causes of differentiation are manifold. They are chiefly the gravitiation field of the earth and the pressure and temperature gradients prevailing during the uprise of the magma. The natural magmatic melt is not only a very hot mixture of metallic oxides in silicate or alumosilicate binding, but represents a complex system in which constituents of low and high volatility are kept in mutual solution. If the system is located at a certain depth in the earth under pressure, it may be in equilibrium but each change in pressure sets off a reaction within the solution phase.

First in the magmatic melt, the precipitation of crystals takes place which, depending on their densities, will sink or rise. This process of material separation is called *crystallization differentiation*. In this, the order of precipitation from an original melt of gabbroid (basaltic) composition is always from minerals which are low in silica to those which are high in silica. This order is given by the so-called Bowen reaction series. Chromite, olivine, pyroxene, magnetite and calcium-plagioclase separate out. Then follow the acid constituents, amphibole and mica, sodic-plagioclase, orthoclase and quartz. The early-formed crystals tend to be dense and sink under gravity whereas the later minerals are lighter and will float upwards in the melt. Because of the early precipitation of the basic components the originally gabbroic melt may acquire a dioritic and, finally, a granitic composition. After the precipitation of the greater part of the silicates there remain only aqueous residual solutions (mostly rich in metals).

This sequence of precipitation demonstrates very strikingly the segregation of elements during differentiation. We can recognize:

(a) Primary crystallization, with minerals rich in Mg, Fe, Ni, Cu, Ti, Cr, Pt.
(b) Main crystallization—producing most igneous rocks composed of Ca, Al, alkalies and Si.
(c) Residual crystallization, showing enrichment of Si, O, S, Cl, F and most metals.

Ore deposits are formed mainly during the Primary and the Residual phases of crystallization.

Besides crystallization differentiation there exists also differentiation in the liquid state—so-called *liquid immiscibility*. This occurs mainly when basic magmas contain larger quantities of sulphides. At high temperatures the sulphides are at first dissolved in the silicate melt. With a low water content of the melt, however (e.g. basic magma with 4% H_2O is regarded as a 'dry melt'), this solubility is mostly so small at temperatures below 1500°, that a separation into two melts takes place (liquid segregation). At these temperatures, neither for the silicates nor for the sulphides of the heavy metals is the point of crystallization reached. With decreasing temperatures the sulphides form droplets increasing in size which sink down by gravity within the magma. There, they may finally gather to form special sulphide melts. The same process takes place in the copper furnace, when the matte (Cu_2S) separates from the slag. In the natural sulphide melts the heavy metals Fe, Ni, Cu, Co are especially involved. The sulphides, accumulated by gravity in the lower zones of the magma body, then solidify within their basic parent rock and crystallize together with the rock minerals, mostly in the form of schlieren or lens-shaped masses (i.e. 'liquid exsolution segregates' analogous to physical metallurgy). These 'simple' processes

1.3. Products of the differentiation of the magma

The natural magmatic melt as already pointed out represents a very complex system, in which constituents of low and high volatility are in mutual solution. The proportion of the constituents of low volatility, which will normally melt only at temperatures well over 1000 °C, amounts to about 90-95% (silicates +oxides). The rest is formed by constituents of high volatility (H_2O, H_2S, HF, HCl, CO, CO_2, chlorides and fluorides of heavy metals etc.). Their melting and boiling points are considerably lower than the melting temperatures of the constituents of low volatility. Therefore, the constituents of high volatility are dissolved in the melt of the low volatile, silicate components. This is, however, only possible if the system is at a certain depth within the earth, that is, if it is subjected to a high external pressure.

It has been found by experiment that the solubility of the constituents of high volatility in the melt is possible only to about 8%. Goranson (1936) determined the solubility of water in a silicate melt by several experiments (Fig. 4). The chief results of these experiments are as follows:

(*a*) In a granitic melt the solubility of water increases with increasing pressure.

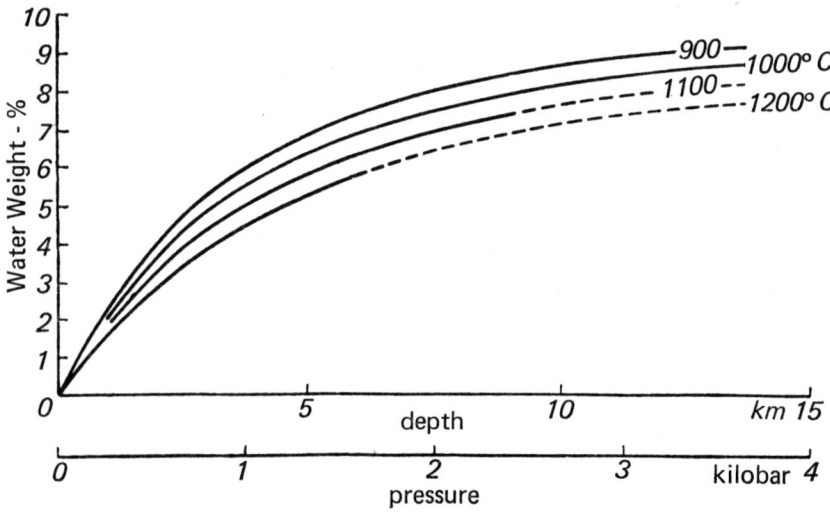

FIG. 4. Solubility of H_2O in albite melts at different temperatures and pressures (after Goranson, 1936).

(b) With rising temperature, however, the solubility of water decreases (thus, a fall in temperature causes the water content of a melt to rise, simultaneously the vapour pressure rises).

(c) With rising water content the melting point of the granite is lowered.

(d) Magmatic melts with a water content of much over 8% are impossible.

If the concentration of 8% is surpassed, that portion of high volatility will separate out. It then forms a second system beside the residual melt. The combination of these heterogeneous constituents of high and low volatility gives the magma, on cooling, its special properties. It behaves like hydrous mixtures which may be fractionated by distillation. On further cooling of the magma and due to the crystallization of the constituents of low volatility, the percentage of the constituents of high volatility (with a low boiling point) increases. Since, however, not all constituents of high volatility can be incorporated into the growing crystals of the constituents of low volatility (for example, OH or F into mica, amphibole etc.), they (especially water) will concentrate more and more in the melt. Because the melt can receive the constituents of high volatility only to a limited extent (water about 8%) these begin to separate out and migrate. Due to the high vapour and gas pressure of the highly volatile constituents the melt acquires a high interior pressure. With the increase of the internal pressure, rising with falling temperature, and the release of the external pressure acting on the system (fissuring in the roof etc.), the easily volatilized constituents are boiled or distilled off. The released constituents of high volatility diffuse through the residual melt in the direction of the pressure gradient upwards, where they gather either in the upwarpings of the pluton roof or migrate through intergranular spaces and fissures in the already solidified top zone. The constituents of high volatility (i.e. residual liquids) are in their turn capable of containing in solution constituents of low volatility (for example, SiO_2, metals) and of transporting them.

These specific properties of the magmatic melt are of decisive importance for the concentration of elements and especially for the formation of ore deposits. It is first of all the separating out of the constituents of high volatility, inevitably associated with the crystallization of a melt, in the form of metalliferous residual liquids that results in a succession of mineral and ore deposits with decreasing pressures and temperatures. The setting free of residual solutions, which takes place in fractions, may go on long after the solidification of the rock minerals of low volatility and we have, therefore, what Niggli (1952) has called the 'differentiation of residual solutions'.

These somewhat complicated processes are very clearly demonstrated on the basis of a simplified binary system, consisting of a constituent of high volatility (A) and one of low volatility (B), in the so-called Niggli diagram (Fig. 5). The *t-x*-diagram shows the saturation curve of B with A for the temperature range from about 1000° to 50 °C. For this temperature range the curve is thus the crystallization curve of B. Assuming the original composition of the melt to be 10% A and 90% B (initial concentration x_1) the crystallization of the melt will begin at the temperature t_1 at point *a*. Due to the increasing segregation of B the residual melt will change in composition in favour of a constantly rising concentration of A. In the area from *a* to b_1 of the curve, at first only the constituents of low volatility of B (silicates) will segregate. That is the stage of the formation of igneous rock including the corresponding liquid magmatic deposits. Between b_1 and b_2 the curve of segregation is almost horizontal, i.e. within a small temperature interval t_2-t_3 the composition of the residual melt changes very strongly from x_2 to x_3 (the increase of the highly volatile phase A results in a considerable rise of the internal pressure *p*; see *t-p*-diagram). In this stage, besides the silicate rocks rich in B, more and more A-containing minerals are formed

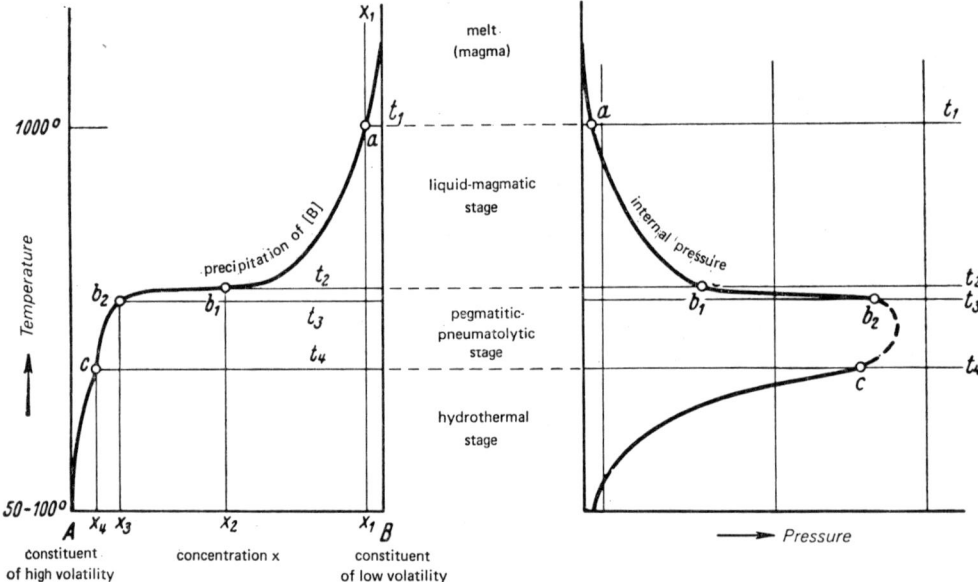

FIG. 5. Temperature-concentration- (left) and temperature-pressure-diagram (right) of a system, consisting of a constituent of high volatility (*A*) and one of low volatility (*B*). (After P. Niggli, from Schneiderhöhn, 1941.) The TX-diagram shows the curve according to which the constituent of low volatility (*B*) separates out with decreasing temperature and simultaneous concentration of that of high volatility (*A*). The TP-diagram shows the initial increase in internal pressure connected with this, up to the pegmatitic maximum value and its further decrease with subsequent condensation.

(pegmatite stage with segregations mainly dependent on pressure).

With decreasing B-content ($b_2 - c$) an increase in the segregation of A takes place (above the critical temperature t_4 =pneumatolytic stage). In the vertical area of the curve below c (below t_2) the segregation of the highly volatile constituents of A finally predominates and these mainly dependent on temperature (hydrothermal stage).

1.3.1. Liquid-magmatic (intramagmatic) formation of deposits

(*a*) DURING PRIMARY CRYSTALLIZATION OF THE MAGMA

The liquid-magmatic stage (1100-600 °C) is mainly determined by the bulk of the silicate melt (=constituents of low volatility). The constituents separated out in the course of primary crystallization can form ore deposits only in those cases where a spatial separation from melt also takes place. With the minerals having a high specific gravity this is mostly effected by gravity (⟶ gravitative crystallization differentiation and liquid immiscibility). Thus the formation of the liquid-magmatic or intramagmatic deposits coincides, as regards time, locality and causes, with the primary crystallization of intrusive (basaltic) magmas. During a period lasting from hundreds of thousands to some million years ultrabasic and basic plutonic rocks differentiate from these magmas and are largely developed in a regular stratiform arrangement.

The main rock groups are:

Ultrabasic plutonic rocks
    Peridotites (predominantly olivine) containing Cr, Pt
    Pyroxenites (predominantly pyroxenes) containing Ti, Fe

Basic plutonic rocks
    Norite } (pyroxenes   containing Ni, Fe, Cu, Co, P
    Gabbro } +plagioclase)
    Anorthosite (predominantly plagioclase) } containing Ti, Fe

Beside the main rock-forming elements (Mg, Fe, Ca, Al, Si, O) the ultrabasic to basic rocks also contain a lot of valuable metalliferous constituents, which can concentrate in certain areas to form ore deposits. These are first of all the siderophilic elements Cr, Ti, Fe (in the ultrabasic rocks) as well as Ni, Fe, Cu, S, Ti and Fe (in the basic rocks). Furthermore, Pt, Co, V, Mn, C and P are also characteristic and occur in smaller quantities or as trace elements.

The concentration of these metalliferous minerals, originally more or less evenly distributed over the basic magma, takes place 'intramagmatically' due to more metals being contained in certain parts of the

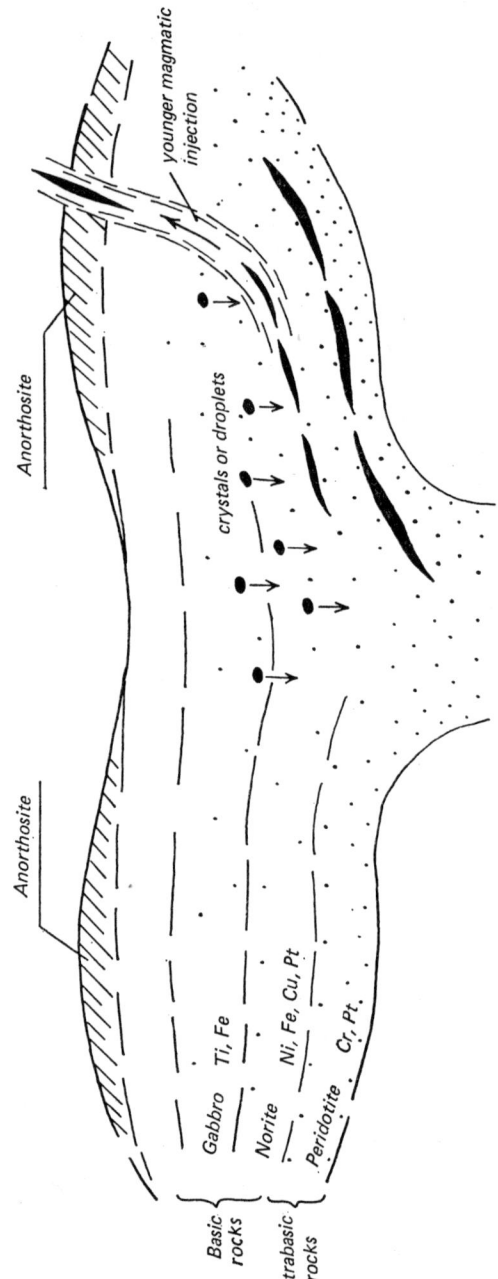

Fig. 6. Schematic intrusion of a basic magma and resultant rocks produced during cooling.

magma from the very beginning or due to subsequent concentration. Separation and concentration may come about by both gravitational crystallization differentiation (⟶ primary segregation and separation by gravity) and unmixing in the liquid state (liquid immiscibility⟶ separation of liquid fractions of the melt and their settling by gravity). Chromite and platinum deposits, mainly confined to ultrabasic rocks, are derived from crystallization differentiation. From the liquid melts of sulphides, however, the nickeliferous pyrrhotite and chalcopyrite deposits originate by unmixing. These are mainly confined to basic rocks and may contain platinum and palladium together with the greater part of the titaniferous magnetite and ilmenite deposits. Frequently such 'liquid-ore magmas', when squeezed out and pressed upward, may form independent, younger injections in the enclosing rock (Fig. 6).

The predominant *structural forms* of the liquid-magmatic deposits are intramagmatic 'schlieren', lenses and bedded deposits. At first the 'schlieren' form cloudy or banded masses of disseminated ores, which become gradually denser to form spherical to ellipsoidal compact lenses and, finally, large deposits of a markedly layered nature (up to several million cubic metres in volume!). Ore-magmas which have been separated by tectonic squeezing also occur as injections in the enclosing rock, frequently in the form of veins and apophyses.

A typical example of such magmatic differentiation resulting in the formation of a series of intramagmatic deposits, is the intrusive complex of the *Bushveld Massif* north of Johannesburg, South Africa. This igneous rock complex forms a gigantic lopolith, which intruded, probably in several stages, into beds of the Upper Algonkian (Precambrian) Transvaal system. Under the weight of the intruded magmas and due to the decrease in volume of the magma beneath, this entire intrusive complex has sagged towards the centre and appears today—after the erosion of the roof—as a vast bowl occupying an area of nearly 100,000 km^2 (450 km west-east, 240 km north-south extension). The vertical thickness of this flat intrusive body amounts to about 9 km. During the slow consolidation of the basic magma a 5 km thick so-called 'Norite series' formed as the lower part of the intrusive complex. It may be assumed that the original magma assimilated considerable amounts of dolomitic limestones from the bottom of the Transvaal system, thus bringing about the formation of the more basic norites. Due to differentiation, this basic rock series shows marked 'magmatic layering' with layers of rocks that can be traced over several 100 km (pyroxenite, norite, anorthosite). Above this rock series there is, as the upper, more acid part of the intrusive complex, the younger so-called 'red granite' (Fig. 7). Recent investigations have indicated that the

16 INTRODUCTION TO ORE DEPOSITS

Bushveld complex is possibly made up of three individual intrusions.
The most important deposits are associated with the norite: at the
bottom, two levels of chromite ore occur (approximately homogeneous

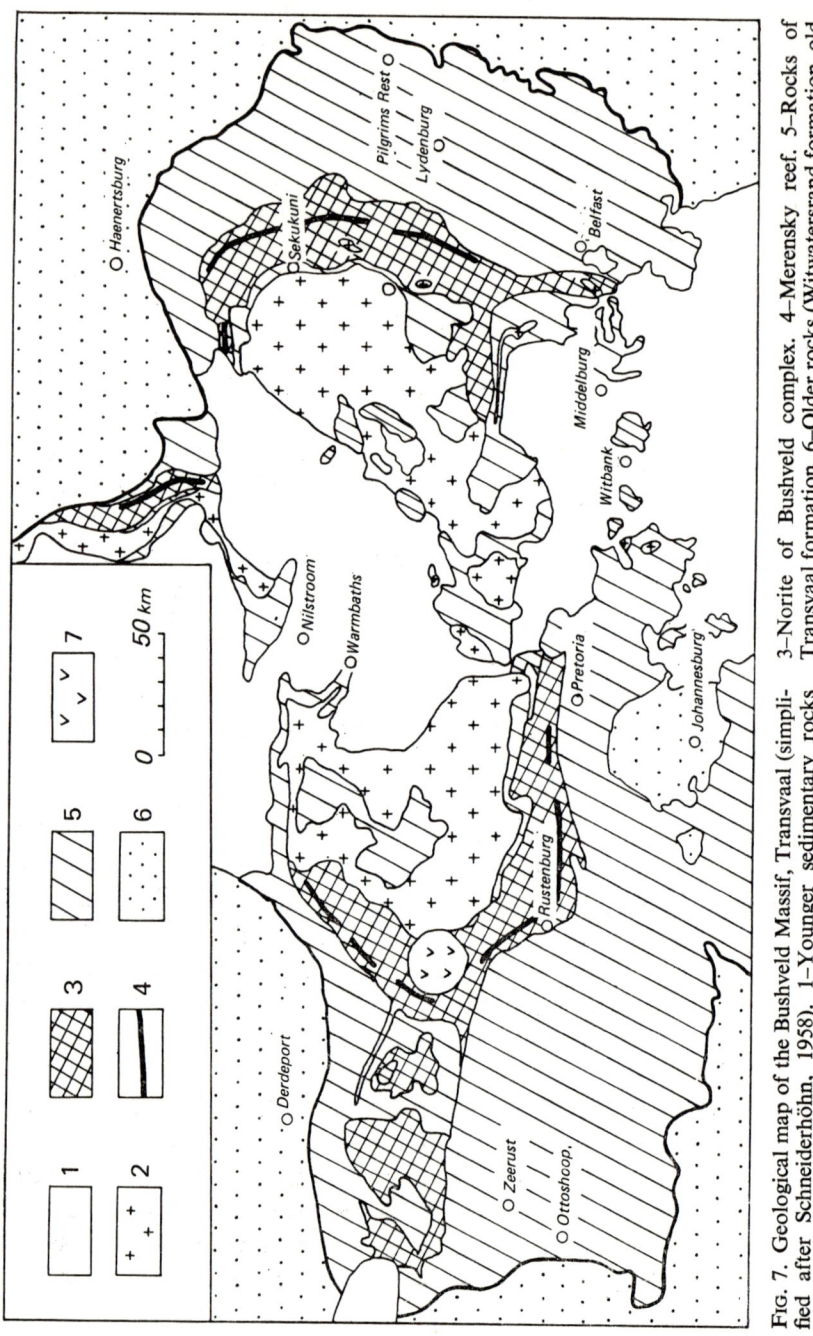

Fig. 7. Geological map of the Bushveld Massif, Transvaal (simplified after Schneiderhöhn, 1958). 1–Younger sedimentary rocks (Waterberg formation etc.). 2–Red granite of Bushveld complex. 3–Norite of Bushveld complex. 4–Merensky reef. 5–Rocks of Transvaal formation. 6–Older rocks (Witwatersrand formation, old granites, gneisses etc.). 7–Younger alkali rocks.

over a distance of 400 km; thickness up to 4 m), then, about 100 m above them, there is a horizon of sulphide ore (the 'Merensky reef' containing (Ni,Fe)S, $CuFeS_2$ and Pt). This ore horizon, which was discovered by the German mining geologist Hans Merensky in 1924, is a conformable layer within the Bushveld norite that can be traced over hundreds of kilometres (thickness up to 9 m). Finally, there occur in the upper parts of the norite two horizons (average thickness =2 m) of titaniferous magnetite (Fe_3O_4 + $FeTiO_3$). These also represent products of liquid-magmatic differentiation of the norite magma. Moreover, it may be mentioned that in the upper parts of the 'red granite', pegmatitic-aplitic zones of gaseous transfer occur, to which tin-ore deposits are bound. Most of the ore horizons or reefs have originated *in situ* by gravitative differentiation (Fig. 8).

m			
5000		Granophyre and felsite of Rooiberg layers	
		Weakly bedded norites, diorites and syenites	Overlying zone
4000		Titanomagnetite	
		Titanomagnetite	
3000		Little differentiated bulky norite	Main norite zone
2000			
		Merensky reef / Chromite reef / Chromite reef — Alternating thin beds of anorthosite, norite, pyroxenite and bronzitite	Differentiated zone
1000			
		Alternating beds of norite, pyroxenite and bronzitite	Transition zone
0		Fine-grained gabbro	Basal zone
		Contact-metamorphosed schists and diabases of the Magaliesberg layers	

FIG. 8. Cross-section of the intrusive complex of Bushveld (from Schneiderhöhn, 1958).

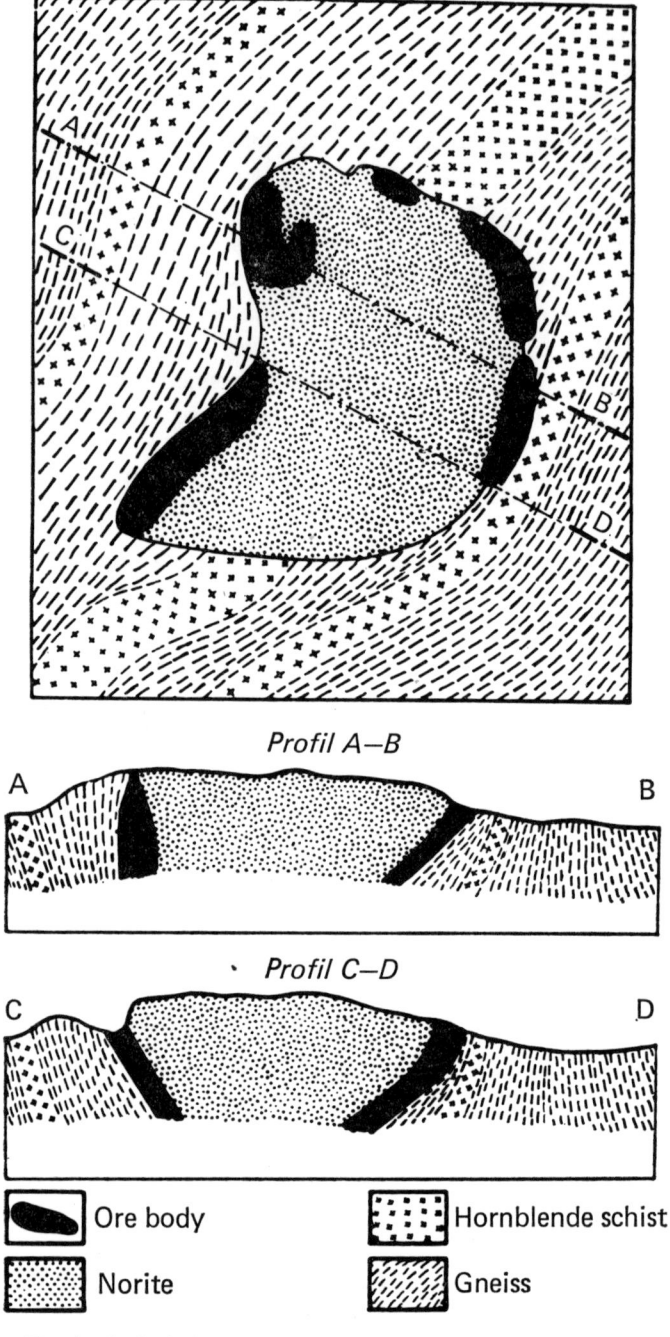

Fig. 9. Geological map and cross-sections of Meinkjär, Norway (after J. H. L. Vogt, 1893).

A further example of the formation of such intramagmatic deposits are the classic nickeliferous pyrrhotite occurrences of Norway. These occurrences, which were formerly of great importance, are located within norites and gabbros. The ores (nickeliferous pyrrhotite +chalcopyrite) are mainly concentrated at the margins and the bottom of the intrusive body (i.e. bottom settlings). The best-known occurrence is that of Meinkjär, southwest Norway (Fig. 9). The ore bodies here always form a sharp contact with the intruded rock, whereas with the norite or the gabbro the boundary of mineralization is gradual and indistinct. The great economic geologist J. H. L. Vogt (1893) investigated these Norwegian occurrences in detail and was the first to develop the concept of the 'liquid-magmatic type of deposits'.

Economically the most important deposit of nickeliferous pyrrhotite in the world is that of *Sudbury*, Ontario. Before World War II 80% of the whole world production of Ni came from here, as well as about 30% of the world production of Pt. Much like the Bushveld Complex, a large bowl-shaped, Precambrian intrusive body (60 × 30 km) lies here between Archean and Algonkian sediments (Upper Huronian). On cooling, the magma differentiated in place, the norite proper forming at the bottom and changing into a micro-granite towards the top zone. At the base of the norite, a bottom layer of nickeliferous pyrrhotite accumulated (Fig. 10). These lens-shaped, compact ore masses at the base of the intrusive body are designated as 'marginal deposits'. In addition, younger, vein-like sulphide deposits were formed ('offset deposits'). These are probably tectonically squeezed-off parts of the main mineralization, which are in part of hydrothermal nature due to their higher contents of volatile constituents (increased occurrence of chalcopyrite and arsenides of Ni and Co). At present these 'offsets' are still the main deposits being mined. The ore being won contains about 3% Ni and 1·5% Cu. The proved reserves still exceed 200 million tons.

(*b*) DURING THE MAIN CRYSTALLIZATION OF THE MAGMA

The main crystallization following the primary crystallization mostly does not supply any ore deposits. In it chiefly the intermediate to acid igneous rocks are formed containing especially the lithophilic elements Si, Al, Ca, Mg, Fe, Na and K. The amount of the common rock-forming silicates is so great that the few ore minerals which separate at this time cannot play an important role (zircon and rutile, for example, are accessory minerals).

1.3.2. Pegmatitic-pneumatolytic formation of deposits

Intensive concentration of the highly volatile constituents is the characteristic feature of the pegmatitic-pneumatolytic phase. This development can only take place after the crystallization of the bulk of the low-volatile silicate mass. The highly volatile constituents are at first present in the supercritical condition (i.e. fluid phase). The term 'supercritical' means here that no difference can be made out as yet between liquid and gaseous states. The substances being in the supercritical condition possess properties characteristic of both liquids and

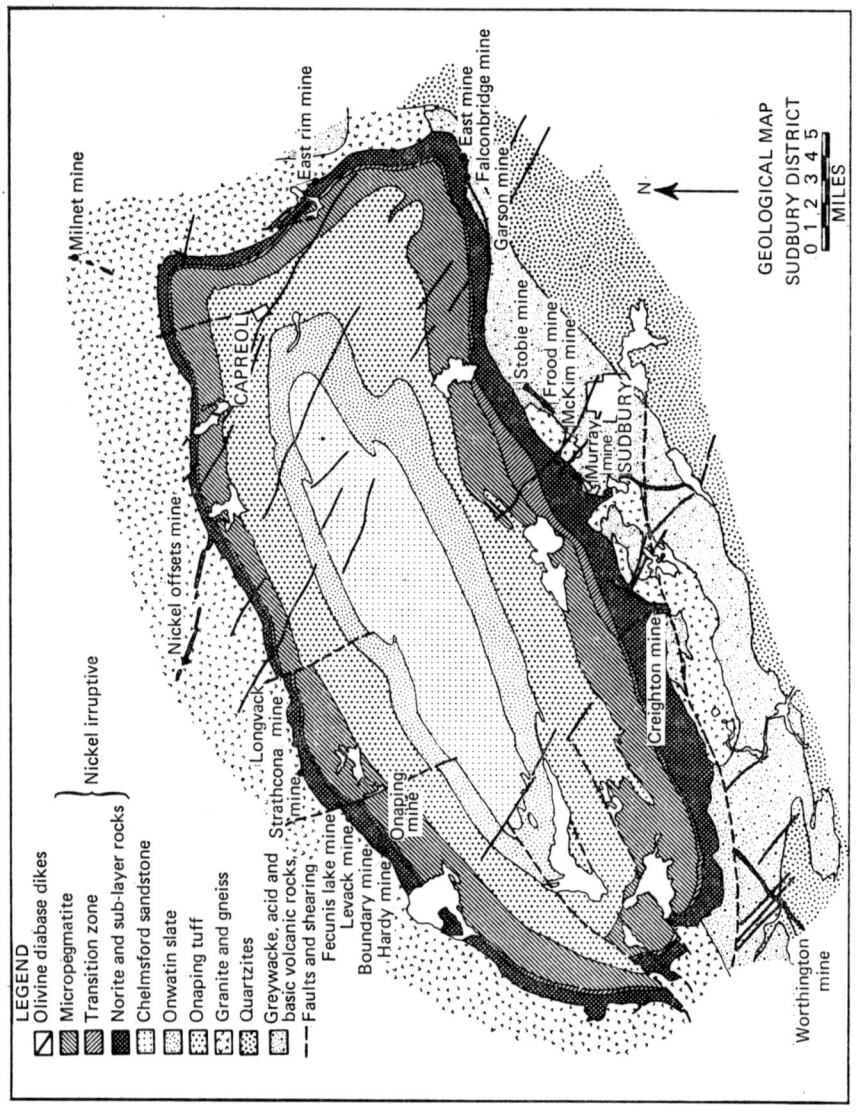

Fig. 10. Geological map of the intrusive body of Sudbury, Canada (from Mertie, 1969).

gases (600-400°C). It is from these supercritical residual solutions, which are especially susceptible to pressure and at first still relatively rich in silicates, that the pegmatitic and pneumatolytic minerals are formed. Here, with the least relief of pressure, minerals are precipitated from the solutions. Such relief of pressure takes place, for instance, when the solutions migrate from the intergranular spaces (i.e. interstices between the individual grains of a rock) into a fissure cavity.

In the *pegmatitic phase* the first substances precipitated from the residual liquids are those in which the liquids are saturated. Therefore the pegmatitic crystals are still similar in composition (mica, feldspars and quartz) to the main magmatic phase of crystallization. Besides these 'rock minerals' there occur also, as accessories, the pegmatite minerals proper. These are mainly formed from those elements (cations) that could not be incorporated into the rock-forming minerals of the igneous rocks because of their ionic radii. For instance, Li, Be and B have small, whereas Rb, Cs, Ba and Sr possess very large ionic radii. Furthermore, elements such as Nb, Ta, Th, Zr, Hf are highly concentrated at this stage. The pegmatites are therefore especially important as deposits of certain nonmetallics (mica, feldspar, quartz, i.e. piezoquartz, apatite) as well as of precious stones (beryl, topaz, tourmaline) and rare earths (monazite, cerite). The typical ore deposits, for example of Li, Al, Sn, W, Mo, U, Au, are found only rarely.

Spatially and genetically, the pegmatites are predominantly associated with acid granitic magmas. All pegmatites lie within the range of the contact aureoles. They are especially frequent in the upper parts of plutons. As to their shape the pegmatites can occur as sharply delimited cavity fillings or as products of metasomatic replacement. In the first case fissure cavities in the marginal zones or in the roof of the pluton are filled ('stockscheider' zone, veins, in part parallel to the contact). In contrast, the metasomatic pegmatites mostly form indistinct 'schlieren' and zones of pegmatization within the uppermost upwarpings of the pluton.

The first-formed pegmatite minerals are distinguished by extremely big grains (single crystals reaching lengths up to 15 m). There are two main causes for this:

(*a*) the very low viscosity and extremely high mobility of the pegmatite solutions;

(*b*) the lack of crystal nuclei and the slow increase in the number of nuclei during growth due to the small energy gradients.

Because of the way in which they form the structure of the pegmatitic deposits is generally extraordinarily irregular, and therefore their exploration and exploitation are very troublesome. As an example of a pegmatite deposit that of *Hagendorf/Oberpfalz* (W. Germany) may be cited (Fig. 11).

Deposits of the *pneumatolitic phase* are formed adjacent to and immediately following those of the pegmatitic phase. They are associated with acid granitic differentiation products. Physically, there is no difference between pegmatitic and pneumatolytic solutions. Both are fluid solutions, whose decisive action, however, depends on their chemistry, that is, ultimately on their parent magma. If the concentration of halogens is relatively low, then the pegmatitic solution (omitting the pneumatolytic stage) may directly change to a hydrothermal solution. Probably, the formation of pneumatolytic solutions

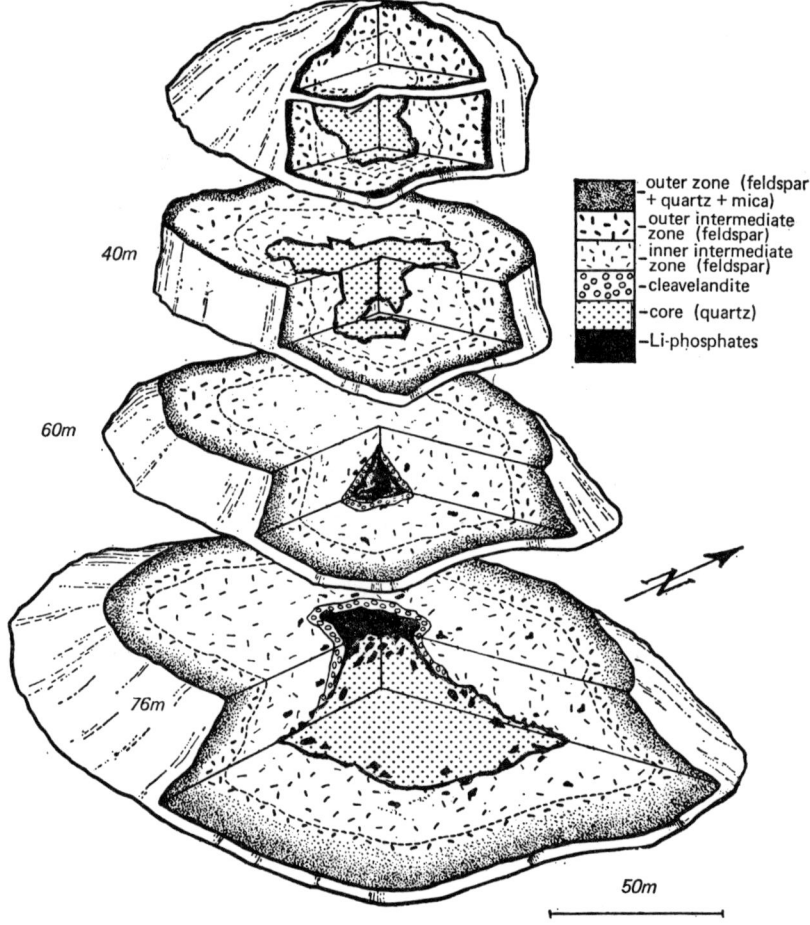

FIG. 11. Block diagram of Hagendorf, Oberpfalz, DDR (after Schneiderhöhn, 1961). A zonally arranged upwarp; the feldspar decreases with depth.

in larger volumes takes place from very acid magmas only. Whereas pegmatitic solutions are still feldspar-forming, the pneumatolytic solutions do not contain any feldspathic material. The feldspars previously formed during the main crystallization and the pegmatitic stage are redissolved and replaced by quartz, topaz, tourmaline, mica and other minerals.*

Pneumatolytic solutions are very active chemically. Water is the main constituent of these solutions. The other highly volatile constituents (primarily fluorides and chlorides of Si and the heavy metals) are mutually dissolved with the water in the supercritical fluid state. In this supercritical state the constituents respond extremely readily to the least changes in the physical conditions (especially P). Then precipitation of substances of low solubility like quartz, cassiterite and others takes place. These precipitations may be thought of as taking place by the following reaction equations:

$$SnF_4 + 2H_2O \rightleftharpoons SnO_2 + 4HF$$
$$SiF_4 + 2H_2O \rightleftharpoons SiO_2 + 4HF \text{ (after Daubré)}.$$

The strong acids released by these processes during ore formation cause strong pneumatolytic transformations in the enclosing rock. Feldspars are especially affected and minerals such as topaz, tourmaline, fluorite and mica are formed.

This classic theory of formation after Daubré is disputed by Barsukov (1957), who holds that Sn is present in the solutions as a complex compound, namely of the type $Na_2[Sn(OH, F)_6]$, which is easily soluble. This compound disintegrates by hydrolysis to form the hydrate of stannic acid, hydrofluoric acid and NaF:

$$Na_2[Sn(OH,F)_6] \rightarrow \underbrace{Sn(OH)_4}_{SnO_2 + 2H_2O} + 2\,NaF + 2HF$$

During the pneumatolytic stage tin-ore deposits are formed together with deposits of tungstenite and molybdenite as well as numerous occurrences of magnetite and hematite. In addition to these elements (Sn, W, Mo, Fe) Bi, Cu and Au may occur in lesser quantities.

As the pressure of the fluid solutions at a short distance from the place of origin decreases so quickly that practically all the least soluble substances must be separated out, the pneumatolytic deposits possess zones of mineralization of relatively small widths (several 100 m).

As regards the *shape of the deposits* (structure), *cavity-fillings* are very common (e.g. veins). On the other hand, because of the great mobility

* 'Greisen' is the name given to the rock in which these alteration products have formed.

of the fluid solutions, mineralization of fissures, impregnations and metasomatic replacement bodies are very common. In the *silicate rocks* (granite, pegmatite, crystalline schist etc.) this results in the formation of the characteristic stock-shaped '*greisen*' bodies. In greisenization the alkalies and alkaline earth salts especially are carried off. As a rule, the amount of the substance removed is larger than that of the substance supplied. Therefore, the greisen bodies reveal a variable but relatively large porosity and thus—if we have not to deal with pure topaz greisen—a *smaller density* than the granite which mostly forms the enclosing rock. This phenomenon is of importance in geophysical exploration using gravimetry.

If, however, the active pneumatolytic liquids come into contact with rocks consisting of *limestone* and *dolomite*, *contact-pneumatolytic replacement deposits* will form. These metasomatic bodies lie mainly outside the intrusive body, mostly in the immediate vicinity of the contact with the calcareous country rock. Layered and concentrically arranged rock zones with contact silicates form from limestone (Fig. 12). When these hard calc-silicate rocks (containing epidote, pyroxene, garnet, amphibole, wollastonite etc.) occur together with ore minerals, they are frequently designated by an old Swedish miners' expression—'skarn'. The main ore minerals may be cassiterite and molybdenite, while the tungstic acid present in the residual liquids did not result in the formation of wolframite ($(Fe,Mn)WO_4$), but in that of the Ca-tungstate, scheelite ($CaWO_4$).

Iron can also occur in relatively large amounts, mainly as magnetite (mostly Ti-free). In this, the magnetite is frequently transformed into pseudomorphs of hematite known as martite. In addition, a voluminous hydrothermal post-phase with sulphides may also occur. These form transitions to metasomatic hydrothermal deposits.

The *structural forms* of the contact-pneumatolytic deposits are very irregular and various and therefore very difficult to explore and estimate. Also the metal contents and the ratios between the individual metals are as irregular as the shapes.

Examples of pneumatolytic deposits

We shall first examine the Sn-W deposits of the Erzgebirge (Ore Mountains, GDR), which are all confined to the most recent acid granites of the Variscan orogenesis. As an example of a pneumatolytic *vein* deposit that of *Pechtelsgrün*/Vogtland may be mentioned. This deposit is located within a granite and consists of a vein system, about 1·5 km long and at an average 0·5 m thick, steeply dipping and striking from northwest to southeast. The vein system consists of several greisen zones parallel in strike and containing younger wolframite-bearing quartz veins. Other pneumatolytic vein deposits especially with Sn are in **Cornwall**.

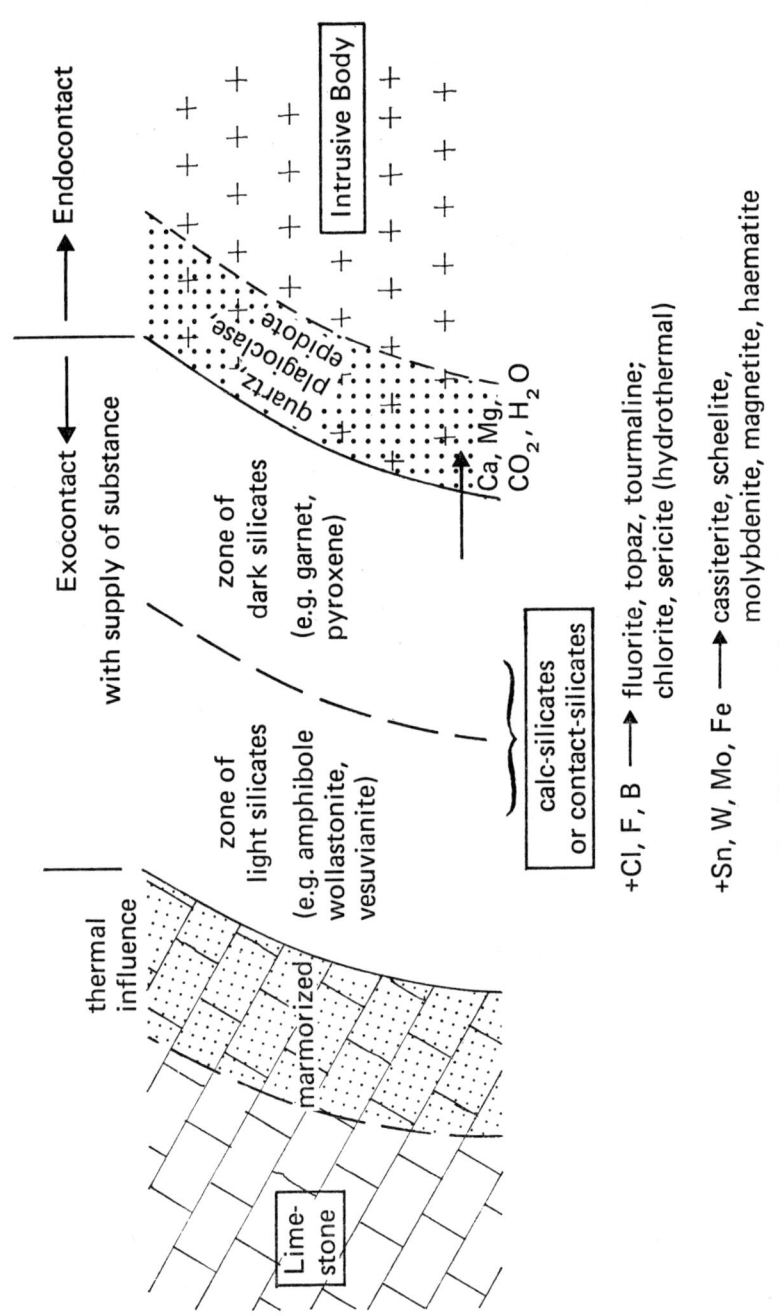

Fig. 12. Pneumatolytic contact zones.

Well-known 'greisen'-bodies occur at *Geyer* and *Altenberg*/Erzgebirge (Fig. 13). One of the *greisen deposits* is also the tin-ore deposit of *Sadisdorf*/Erzgebirge. Here, a conical granite body sits within an older

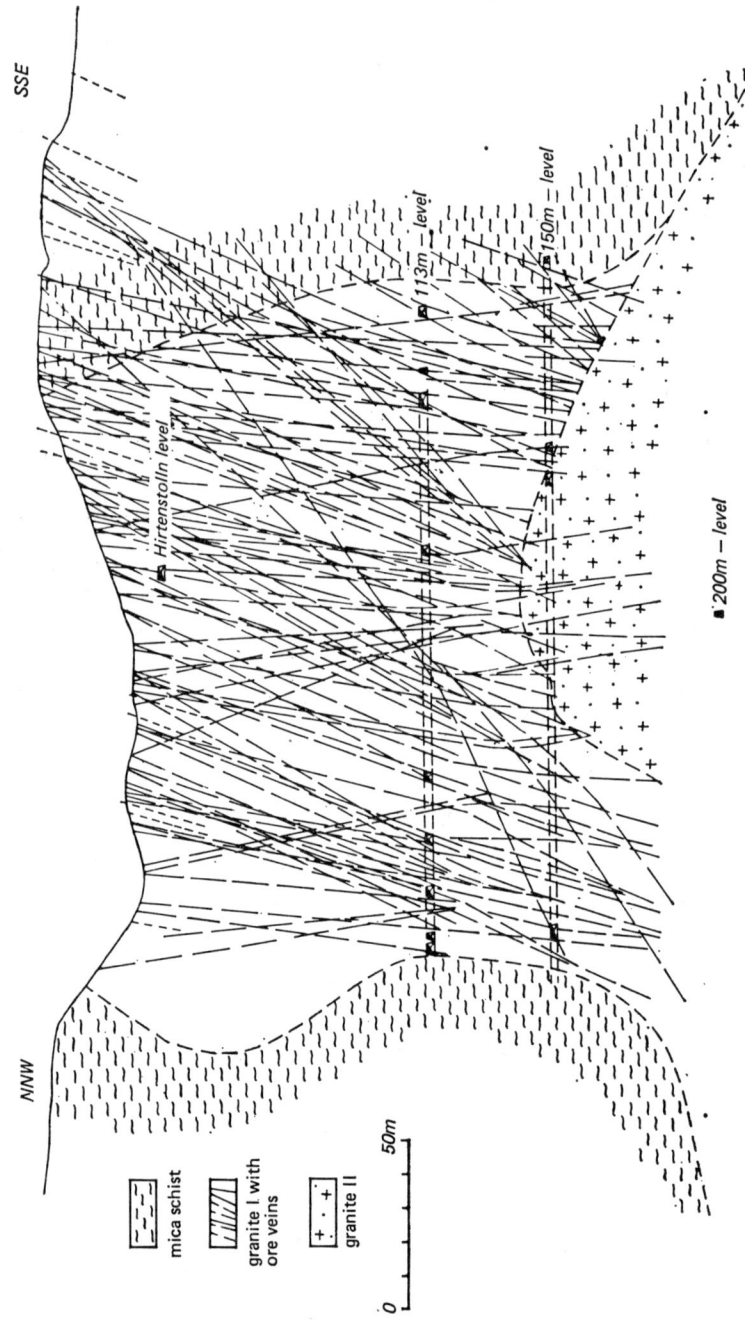

FIG. 13. The 'greisen'-body of Geyer, Erzgebirge (GDR). Numerous fissures with greisenization in the upper parts of the granite upwarp (after Bolduan, 1963).

external granite. With this external granite the internal granite is marked distinctly by a marginal pegmatite ('stockscheider' or stock separator; Fig. 14). The internal granite itself is completely changed into greisen from its top down to a depth of 250 m. The 'stockscheider' too, in its upper parts, is completely transformed to a quartz-mica greisen with giant grains. This so-called 'quartz bell' with its coarse-crystalline fabric and its small intergranular spaces prevented the residual liquids of the internal granite from migrating upwards. Therefore, greisenization is most intensive here. The metallization of the greisen consists of cassiterite, some wolframite, molybdenite and a final sulphide phase.

Examples of *contact-metasomatic* origin are provided by the iron-ore deposits of the *Banat*/Rumania. B. v. Cotta recognized and described this type of deposit for the first time in 1864. Mesozoic limestones,

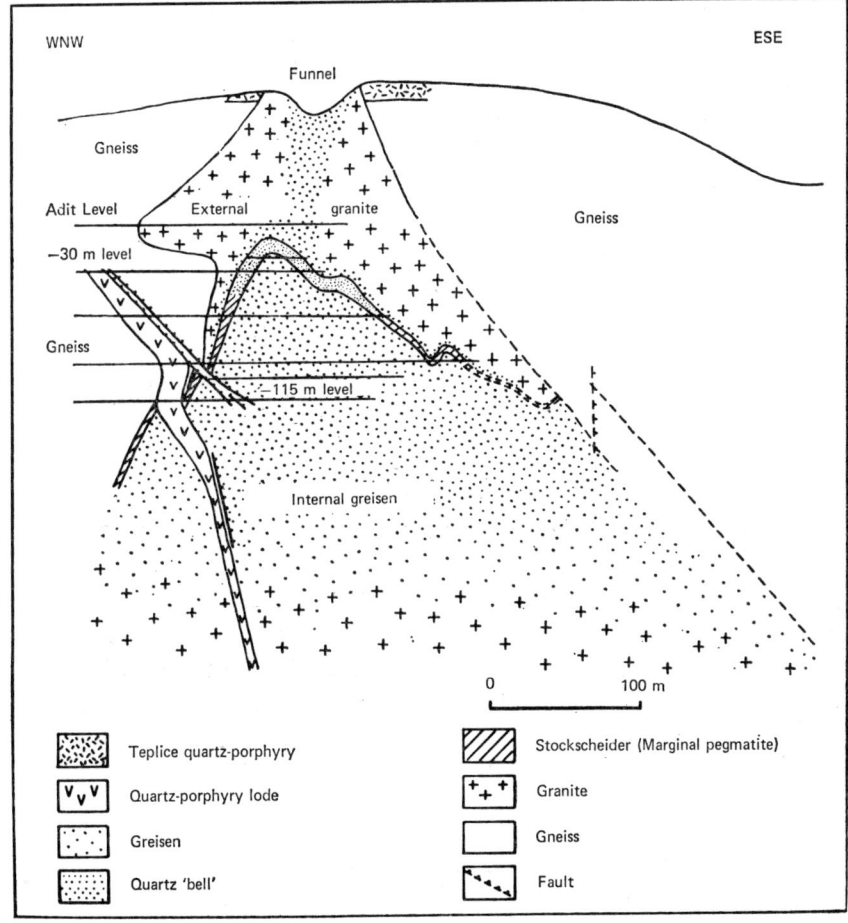

FIG. 14. The tin-greisen deposit of Sadisdorf, Erzgebirge (GDR) (Baumann, 1970). 1970).

embedded in gneiss, are transformed by a Cretaceous granodiorite to skarn and ore (magnetite, some hematite and sulphides; Fig. 15). The following equation for the reaction of the Fe-solutions with the limestone is assumed to apply:

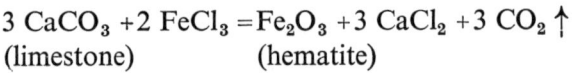
$$3\ CaCO_3 + 2\ FeCl_3 = Fe_2O_3 + 3\ CaCl_2 + 3\ CO_2 \uparrow$$
(limestone)   (hematite)

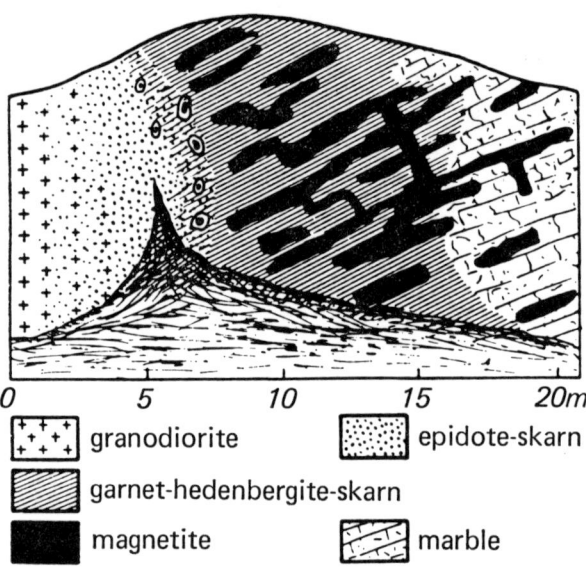

FIG. 15. Open-pit wall of the contact-metasomatic iron-ore deposit of Vaskö, Banat (after Schneiderhöhn, 1941).

1.3.3. Hydrothermal formation of deposits

On their way through the fissure systems the supercritical solutions will reach areas of lower temperature and cool gradually. Finally, the temperature will be below the critical temperature of water and the fluid solutions will condense to become hot aqueous (i.e. hydrothermal) solutions. The critical temperature of pure water is 374·1 °C, but if other substances are present in the solution it increases to about 400 °C. The greater the distance between where a hydrothermal mineral sequence separates out and its magma source, the lower will be its temperature. Accordingly, there are distinguished deposits of:

(a) katathermal origin   (400-300 °C)
(b) mesothermal   ,,   (300-200 °C)
(c) epithermal   ,,   (200-100 °C)
(d) telethermal   ,,   (100- 0 °C)

FORMATION BY MAGMATIC PROCESSES

Whereas the supercritical solutions are still very mobile and can migrate between the intergranular spaces, the aqueous hydrothermal solutions are not capable of this due to their surface tension. Therefore, for the continued transportation of the hydrothermal solutions and the formation of deposits, jointing and cavities are necessary. Tectonic movements resulting in the formation of joints and veins now acquire a decisive importance for the migration and deposition of metals and mineral constituents.

At present nothing is known definitely about the *nature* of the hydrothermal solutions and the mode of transportation of the heavy metals. The aqueous solutions contain dissolved salts such as chlorides, fluorides, sulphates, bicarbonates, sulphides, silicates (that is, the ions Cl^{1-}, F^{1-}, SO_4^{2-}, CO_3^{2-}, S^{2-}, SiO_4^{4-}), alkalis (Na, K), alkaline earth salts (Mg, Ca, Sr, Ba), dissolved gaseous constituents (CO_2, H_2S, SO_2 and so on) as well as a great number of heavy metal ions (Fe, Co, Ni, Mn; Au, Ag, Cu, Zn, Pb, Sb, Hg and so on).

The solutions may be completely ionized, they may be colloidal or they may occur as mixtures of both (i.e. complex solutions). In the supercritical and katathermal states the *transport of the metals* probably takes place in anion-dispersed or molecule-dispersed solution (i.e. true solution). As the solubility of the sulphides is very low in hot waters, the relatively great amounts of metals separated out in the deposits cannot have been transported in the form of pure sulphide solutions (ZnS–PbS– solutions; Fig. 16). Because of this, it is assumed that the heavy-metal bisulphides being more easily soluble play an important part in ionic solutions. American writers (e.g. Barton 1959, Krauskopf 1964) in particular suggest that hydrothermal transport of the elements must take place in complex compounds.

The other possible mode of transportation is in colloidal solution. 'Colloidal solubility' is higher by some orders of magnitude than 'ion solubility', that is, in a colloidal solution much greater amounts can be transported per unit dispersion medium.

	ZnS	PbS	CuS	HgS	Ag$_2$S	Cu$_2$S	
25°C	1.42×10^{-7}	8.65×10^{-9}	2.4×10^{-13}	1.0×10^{-18}	3.8×10^{-13}	4.8×10^{-10}	Cu$_2$Sα
100°C	3.6×10^{-4}	8.9×10^{-8}	4.1×10^{-13}	2.2×10^{-17}	5.7×10^{-13}	4.0×10^{-11}	
200°C	2.2×10^{-2}	2.1×10^{-6}	4.6×10^{-12}	6.4×10^{-16}	3.3×10^{-12}	5.6×10^{-10}	
300°C	0.8	1.6×10^{-5}	2.3×10^{-11}	5.6×10^{-15}	1.4×10^{-11}	2.6×10^{-9}	Cu$_2$Sβ
400°C	5.92	2.1×10^{-4}	7.8×10^{-11}	3.1×10^{-14}	2.3×10^{-11}	7.3×10^{-9}	

Fig. 16. Solubility of sulphides in g/l (after Verhoogen, 1938).

For hydrothermal mineral precipitation the following factors are recognized:

1. Temperature;
2. Pressure (to a minor degree);
3. Material concentration;
4. pH value (concentration of hydrogen ions);
5. Eh value (concentration of oxygen, i.e. redox potential).

These five parameters characterize the physico-chemical conditions of mineral formation, and they differ in importance:

1: The saturation limits of the various materials are reached successively as the solutions decrease in temperature. This results in mineralogical differences related to depth of occurrence.

2: The effect of pressure is small because of the fact that water has a low compressibility.

3-5: The main pre-condition for mineral deposition is a sufficient concentration of the elements. The precipitation of the elements present in the solutions as *ions or molecules* (true solution) depends on the pH and Eh values, as well as on the ability to form insoluble compounds with the still-existing constituents in the system. Thus, some elements (Fe, Mn, Sn) form mainly under conditions of oxidation, as with the increase of their valence the solubility of the corresponding compounds is diminished. On the other hand, the preferential precipitation of other elements (Cu, Pb, Mo) under reducing conditions is explained by the formation of compounds of low solubility at low valence. Then, with given concentration of the anions and making allowance for the Eh and pH values, the ranges of stability (i.e. ranges of precipitation) of the forming ore minerals or mineral sequences (parageneses) can be determined (Fig. 17).

In the supersaturated *colloidal* solutions, the elements are subjected to other controls not yet sufficiently known. The conclusion may be drawn from the very frequent occurrence of colloidal (gel) structures and colloidal minerals that these solutions must have played a very important role in the formation of hydrothermal deposits. The supersaturated solutions form so-called sols (they contain as a disperse phase the minerals themselves), from which the mineral coagulates. This means that it is deposited in the form of a gel to crystallize slowly in the gel mass. The process may take place both at low and high temperatures. The formation of gels indicates only a very high concentration in the solutions. The coagulation to gels and the resulting formation of minerals may be effected by electrolytes (reaction with older ores or country rock), by mutual flocculation of several phases with different

FORMATION BY MAGMATIC PROCESSES 31

electric charge, by alterations of T, P, pH, Eh and so on.

Numerous local modifications exist among the great number of the hydrothermal ore deposits of the world; nevertheless a general scheme of the succession of metals and assemblages can be established. The aqueous hydrothermal solutions consist mainly of highly volatile constituents and a series of chalcophile heavy metals (Fe, Co, Ni, Mn; Au, Cu, Zn, Pb, Zg, Bi, Sb, Hg etc.) mostly bound to sulphur (partly also to As, Sb, Se). The principal kinds of gangue, which form besides quartz (SiO_2), are carbonates (of Fe, Mn, Mg, Ca), sulphates (of Ba, Ca) and fluorite (CaF_2).

Hydrothermal deposits comprise a major part of the ore deposits

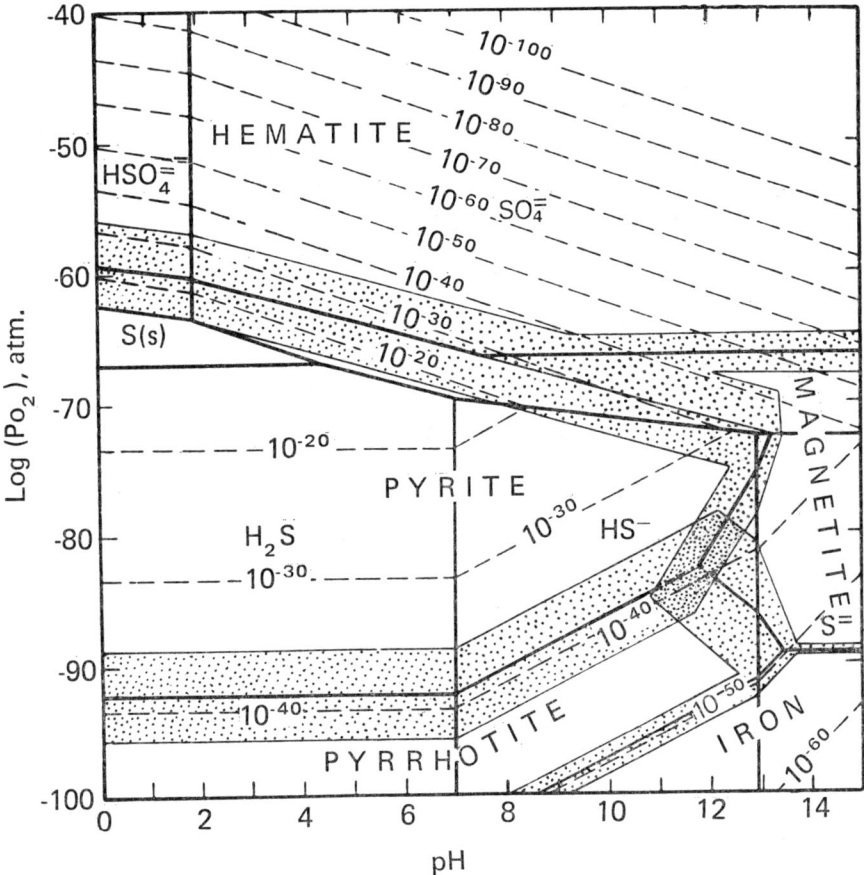

FIG. 17. Equilibrium relations between iron oxides, iron sulphides, and aqueous solutions at $(\Sigma S) = 0.001$ m and 25 °C. Dashed lines are contours of P_{S_2} in atmospheres. Heavy lines bound the areas of predominance of each ion or molecule. Shaded zones indicate the uncertainty in the limits to the fields of stabilities of the various minerals (after Barnes and Kullerud, 1961).

derived from magmatic activity. The various metals occur associated in characteristic mineral parageneses, which can be grouped into definite 'vein-ore formations':

1. Au-Ag formation
2. Fe-Ni formation   } katathermal origin
3. Cu-Fe-As formation
4. Pb-Zn-Ag formation
5. U-Fe formation   } mesothermal origin
6. Fe-Mn-Ba-Fe formation
7. Bi-Co-Ni-Ag formation   } epithermal origin
8. Sb-Hg formation

These ore formations are persistent parageneses, that is they are worldwide with only insignificant variations in mineral composition.

The *structural forms* of hydrothermal deposits are the ore veins or lodes. Due to their tectonic origin the veins mostly reveal a regular parallel arrangement in systems, which are connected with the geologic-tectonic development of the whole district in question.

The mainly two-dimentional, plate-like vein bodies may either penetrate the older rock-layers unconformably or conformably as bedded veins in the stratification of the country rock (Fig. 18). The limiting surface of the veins, that is the vein walls, are termed 'salbands'. The strike-wise extension of ore veins varies from 100 m to several kilometres: (for example: Mother Lode, California = 120 km). The greatest extension downward proved up to now reaches 3000 m. The workable thickness, of prime importance to the miner, depends very much on the value of the ore (with uranium veins, thicknesses are measured in millimetres, with nonferrous metals in decimetres, with

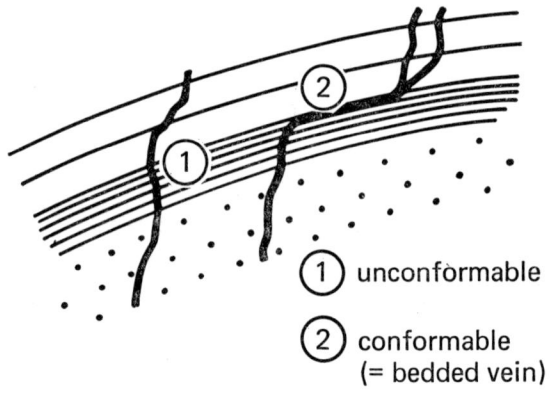

FIG. 18. Unconformable and conformable formation of ore veins.

siderite in metres). The distribution of mineralization within the veins is never uniform. Usually only one-quarter to one-third of the whole vein area is so metallized as to be workable. This portion is expressed by the coefficient of metallization or workability. Ore concentrations in the veins are also designated as ore shoots. They may have developed because of physical causes (broadening of fissures by rock inhomogeneities) or chemical causes (by bituminous, basic or ore-bearing country rocks, intersection of lodes or vein crossings; Fig. 19).

The prototype of hydrothermal deposits is the ancient ore district of Freiberg. The Freiberg ore district occupies an important place within the Erzgebirge metal province. 800 years ago the first silver was found in the area of the present town of Freiberg. This discovery led to the foundation and the subsequent development of the mining town of Freiberg with its famous mining academy. Mining in the Freiberg district greatly influenced the historical, economical and cultural development of the former country of Saxony and indeed had an

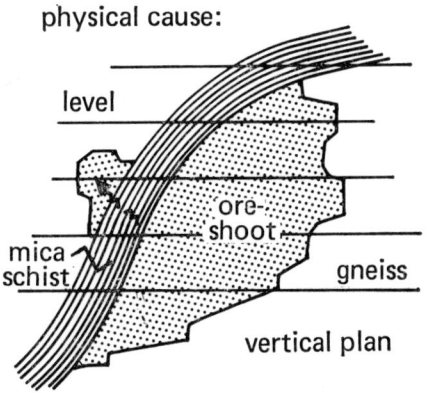

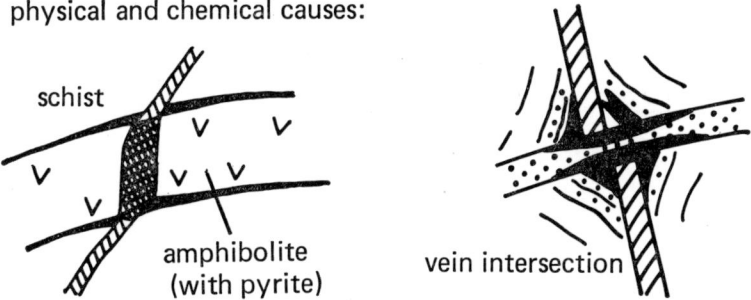

Fig. 19. Some causes of ore-shoots. Physical cause: ore solutions 'dammed-up' due to impermeable rock. Chemical cause: bituminous or sulphide-bearing country rocks as well as older veins aid in precipitation from ore-solutions.

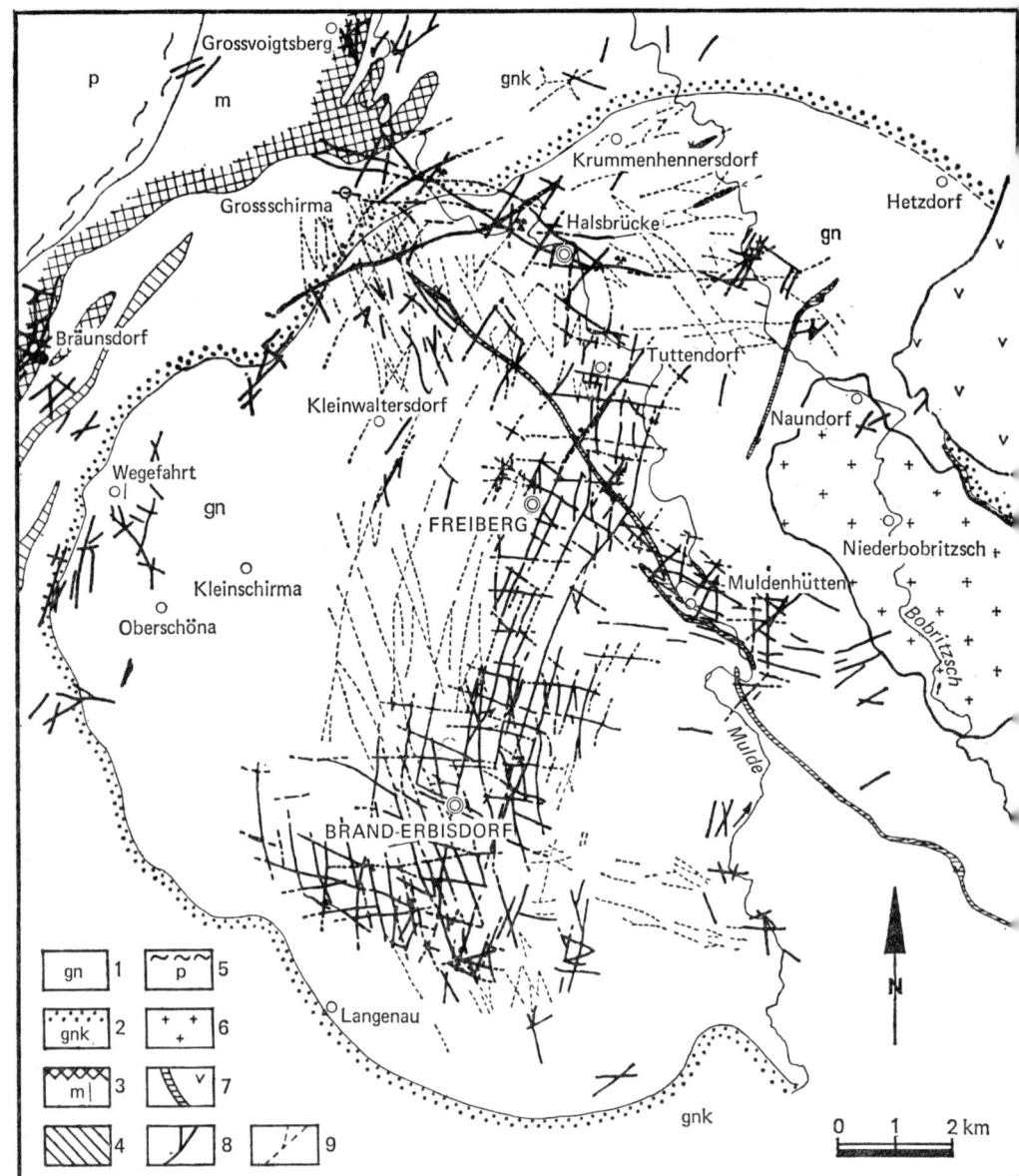

FIG. 20. Geological sketch map of the central part of the Freiberg ore-district (after Baumann, 1965).
  1–biotite,
  2–mica-gneiss,
  3–mica schist,
  4–red gneiss rievaties,
  5–phyllite,
  6–granite,
  7–porphyry,
  8–ore veins,
  9–tectonic structures.

effect on all the mining districts of Central Europe.

The ore veins of the Freiberg ore district are interspersed in the crystalline basement of the eastern Erzgebirge (Fig. 20). This is mainly composed of the gneissose rocks revealing in their foliation a curving strike and a dip increasing from the centre outwards. On the margin of this central gneiss complex there follow, onion-skin-like, further varieties of gneiss. Towards the northwest the gneisses are followed by micaschist, phyllites and Palaeozoic rocks of the Erzgebirge syncline. Within this older rock setting there occur also younger Variscan intrusive bodies (biotite granite, quartz porphyries in the form of nappes and veins).

By intensive mining activity more than 1000 ore veins were opened up in the Freiberg area. Two fissure systems running normal to each other can be distinguished, namely, one vein system trending approximately N–S and one trending approximately E–W. Two types of fissure can also be distinguished, which are interpreted as shear joints and feather joints. The shear joints are distinguished by their great extension along the strike (up to 20 km) and their vertical dip. The thickness of the shear joints is relatively great (up to 6 m), the rock material being mostly strongly crushed. Mineralization on these veins often took place in the form of impregnations. The extension of the feather joints (NW–SE) alone the strike is much smaller (2 km at an average) and they have a dip of 30-70° west. The ores on these veins are characterized by a structure ranging from massive to layered.

Soon after the initiation of mining at Freiberg it was found that the mineral content of the ore veins is by no means uniform. There can be distinguished several 'ore formations', which differ from each other by their characteristic mineral assemblages and, in part, also by the strike of the host veins.

According to recent research the following six ore formations are distinguished (Fig. 21):

I. *Tin-tungsten formation*

Pegmatitic-pneumatolytic sequences are not very common in the Freiberg district but they occur in their typical form only towards the south (area of Altenberg, Fig. 22).

II. *Polymetallic lead-zinc formation*

This consists of three sequences:
 (*a*) Pyritic sequence with quartz, arsenopyrite, pyrite;
 (*b*) Zn-Sn-Cu sequence with sphalerite and subordinated stannite and chalcopyrite; and
 (*c*) Lead sequence with galenite rich in silver.

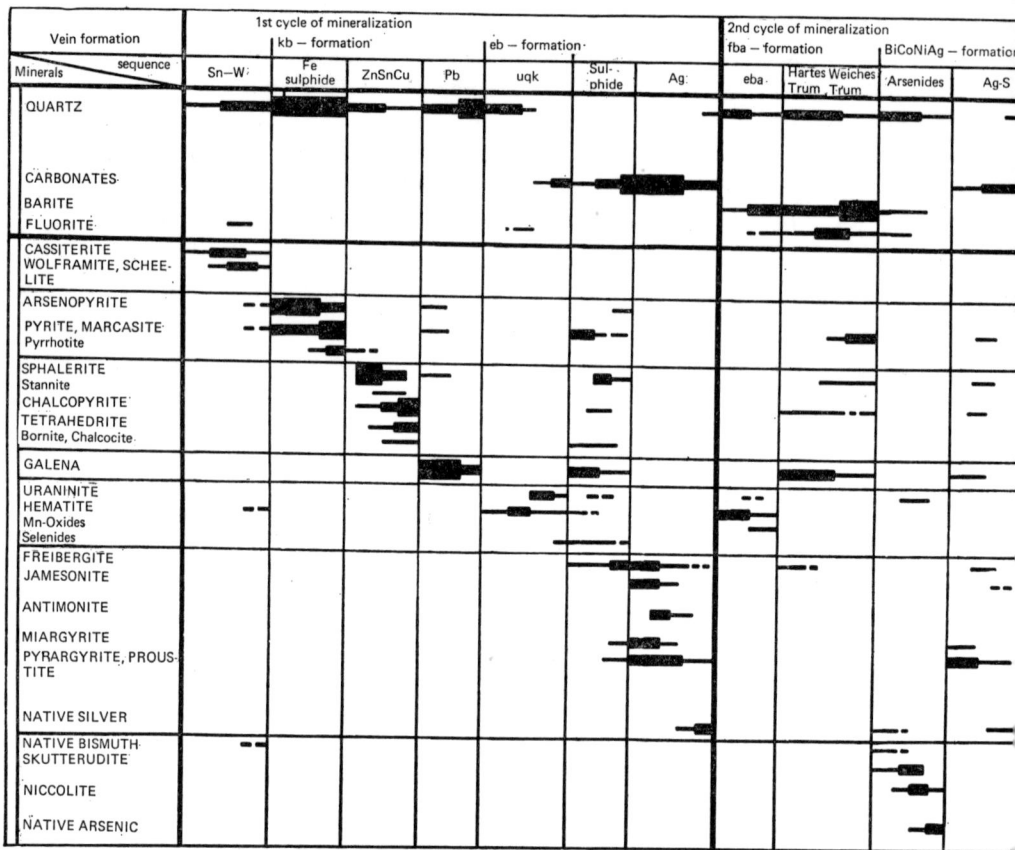

FIG. 21. Mineralization sequence of the ore veins of the Freiberg ore district (modified after Baumann, 1967).
kb = *k*iesig-*b*lenbdige Bleierzformation (= pyritic sphalerite-galenite formation)
eb = *E*dle *B*raunspatformation (= precious carbonate formation)
uqk = *U*ran-*Q*uarz-*K*alzit-Abfolge (= uran-quartz-calcite sequence)
eba = *E*isen-*B*aryt-*A*bfolge (= Iron-baryte sequence)
fba = *F*luorit-*B*aryt-Formation (= fluorite-baryte formation).
1st cycle of mineralization = Variscan.
2nd cycle of mineralization = Post-Variscan.

III. *Precious carbonate formation*

This is also subdivided into three sequences:
- (a) Uranium-quartz sequence with cherty quartz and pitchblende;
- (b) Sulphide carbonate sequence with carbonates, freibergite and re-deposited sulphides from the polymetallic lead-zinc formation;
- (c) Silver-rich sequence with carbonates and manifold silver ores (pyrargyrite, argentite, native silver etc.).

IV. *Fluorite-baritic formation*

This formation reveals also a subdivision, namely into a red-iron-baryte sequence with quartz, baryte and hematite, and a baryte-fluorite-sequence with baryte, fluorite and quartz and the ore minerals galenite, sphalerite and marcasite.

V. *Bi-Co-Ni-Ag formation*

Here an arsenide sequence and a sulphide sequence are distinguished. The arsenide sequence is but poorly developed in the Freiberg ore district. The sulphide sequence contains especially silver minerals besides sulphides (e.g. proustite, argyrodite and others) and occurs mainly at vein intersections.

VI. *Iron-manganese formation*

This is made up of quartz and iron and manganese oxides. They are especially confined to the large NW–SE-trending structures of the Erzgebirge.

According to the succession of mineralization there can be observed in the Freiberg ore veins a lateral change in facies, which can be noticed in a corresponding spatial zonation of the assemblages within the ore district (Fig. 22). Based upon geological, paragenetic and geochemical investigations as well as isotope determinations on some minerals it is possible to clearly distinguish two chronologically different cycles of mineralization, belonging to different orogenic and, consequently, different metallogenic epochs:

(1) The Variscan cycle of mineralization (tin-tungsten, polymetallic and precious carbonate formations);
(2) The post-Variscan cycle of mineralization (fluorite-baritic and Bi-Co-Ni-Ag formations).

While the sequences of the Variscan or Hercynian cycle of mineralization are probably connected in a certain way with palingenetic (remelted) granite intrusions, the sequences of the Variscan or Post-Variscan cycle of mineralization are thought to be the products of simatic magmas.

Hydrothermal deposits do not only occur in vein fissures but frequently also in the form of *metasomatic replacement bodies*. These develop from the contact-pneumatolytic replacement deposits already described. Hydrothermal metasomatism or replacement is a process, in which new mineral aggregations are formed. For the formation of vein deposits, the mechanical properties of the country rock are of great importance whereas the chemical properties of the country rock are the effective controls in the metasomatic deposits. Due to the relatively low

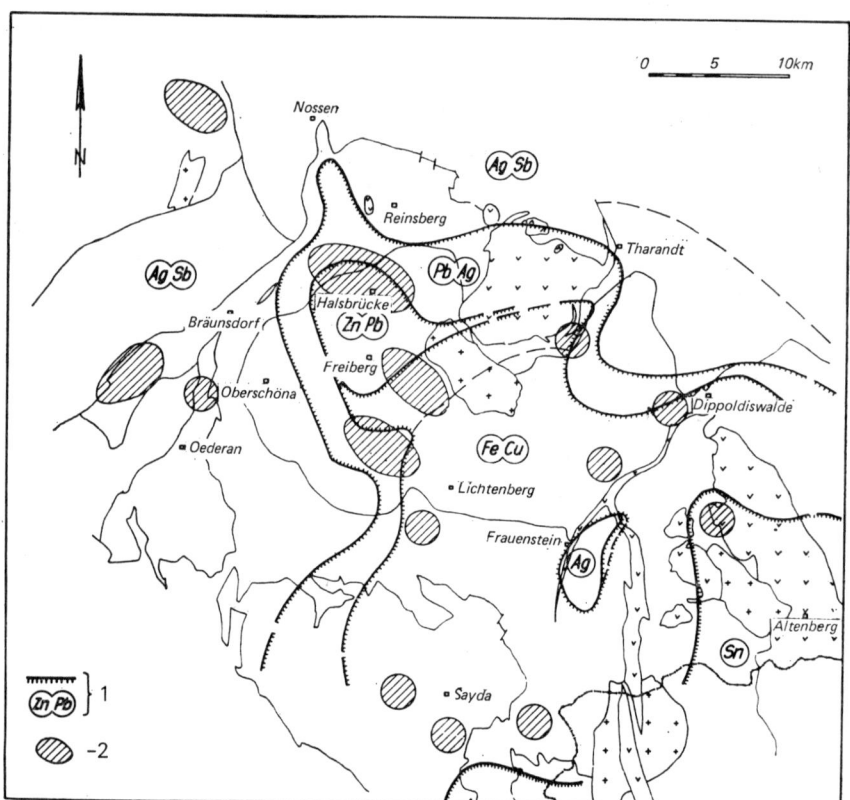

FIG. 22. Zonal arrangement within the Freiberg mining district (after Baumann, 1965).

1 – zonal sequences of the Variscan mineralization cycle,
2 – distribution area of the Post-Variscan mineralization cycle.
For geological situation see Fig. 20.

temperatures of the solutions and the general high reactivity of carbonates the metasomatic replacement deposits in a narrow sense are formed primarily in *carbonate rocks*:

$$CaCO_3 + Fe(HCO_3)_2 = FeCO_3\downarrow + Ca(HCO_3)_2$$
$$\text{(siderite)}$$

Two solid phases and one liquid phase always participate in the process of metasomatism. The solid phases consist of the partially replaced pre-existing material (pelasome) and the newly formed mineral aggregations (metasome or metacryst; Fig. 23). According to the law of mass action the liquid phase causes the reaction between the two solid phases to take place.

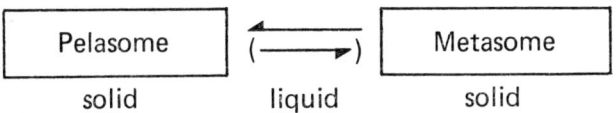

FIG. 23. Process of metasomatic replacement.

The rock horizons impermeable to liquids are of special importance for the location of metasomatic deposits. At such levels the rising solutions are effectively dammed thus effecting intensive replacement, transformation and metallization. The siderite concentrations near *Schmalkalden* in the Thuringian Forest (GDR) provide a clear example of this. They lie in Zechstein limestone under the 'lean' clay occurring there (Fig. 24).

The mineral content in the metasomatic deposits is similar to that of the veins, with the difference, however, that it is still more mixed as it consists both of the newly supplied materials and of the partly transformed remains of the replaced country rock.

One of the largest metasomatic siderite deposits is the *Ore Mountain* (Erzberg) in Steiermark (Austria). Here, a thick block consisting of Devonian limestone has been changed to siderite. The lower wall consists of schistose porphyry rocks. The upper wall is composed of the Werfener beds (Lower Triassic). The latter exerted, it seems, a certain damming effect (Fig. 25). In this way gigantic compact masses of siderite were formed, approximately unchanged in quality over a vertical distance of over 1000 m. The ore being won contains about 35% Fe and 2·5% Mn. The reserves at present still amount to about 400 million tons. The output is approximately 1·5-2·0 million tons per year. The ore is won in a gigantic open-cast mine (70 levels) and by underground mining.

Another important metasomatic siderite deposit is located farther south near *Huettenberg* in Austria (in Cambrian marbles). Further occurrences–in part also with magnesite—are located in Slovakia and in the Balkans (Bosnia, Banat). These occurrences belong genetically to the large unit of the Alpine-Carpathian orogenic system. This also applies probably to the partly hematite-bearing occurrences in North Africa (Morocco, Algeria, Tunisia, in Lias and Lower Cretaceous limestones). An important metasomatic iron deposit is that of *Bilbao*, North Spain. This is made up of metasomatic siderite which has been

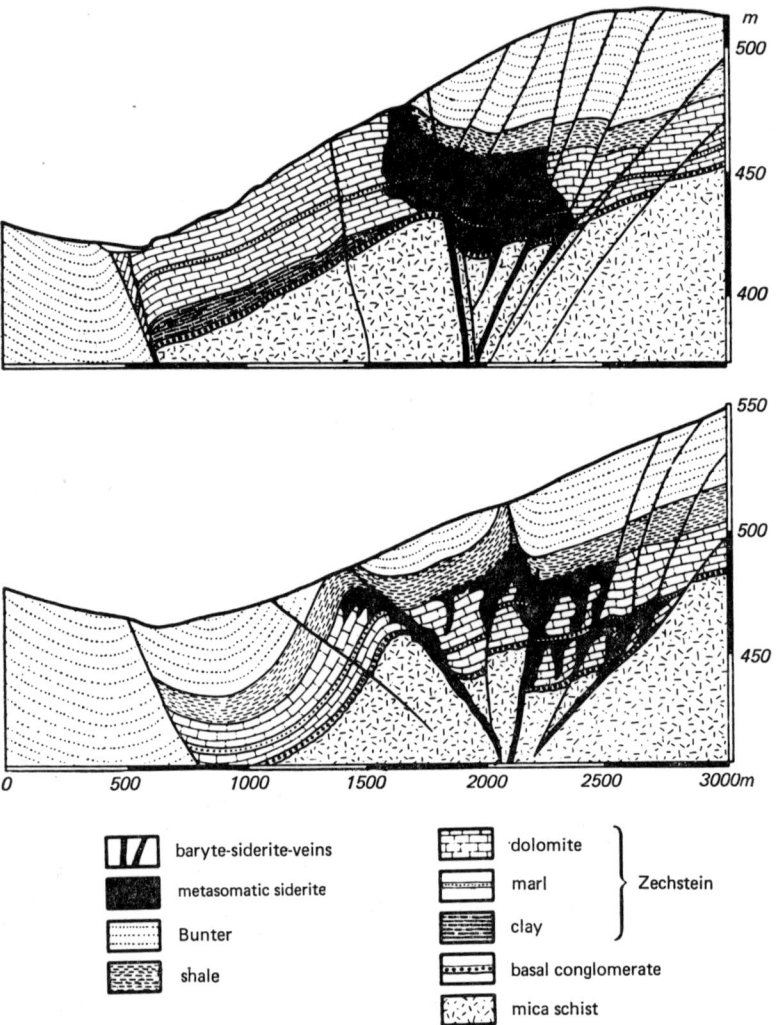

FIG. 24. Two cross-sections of the ore-deposit of Schmalkalden, Thuringian Forest (DDR) (from Schneiderhöhn, 1941).

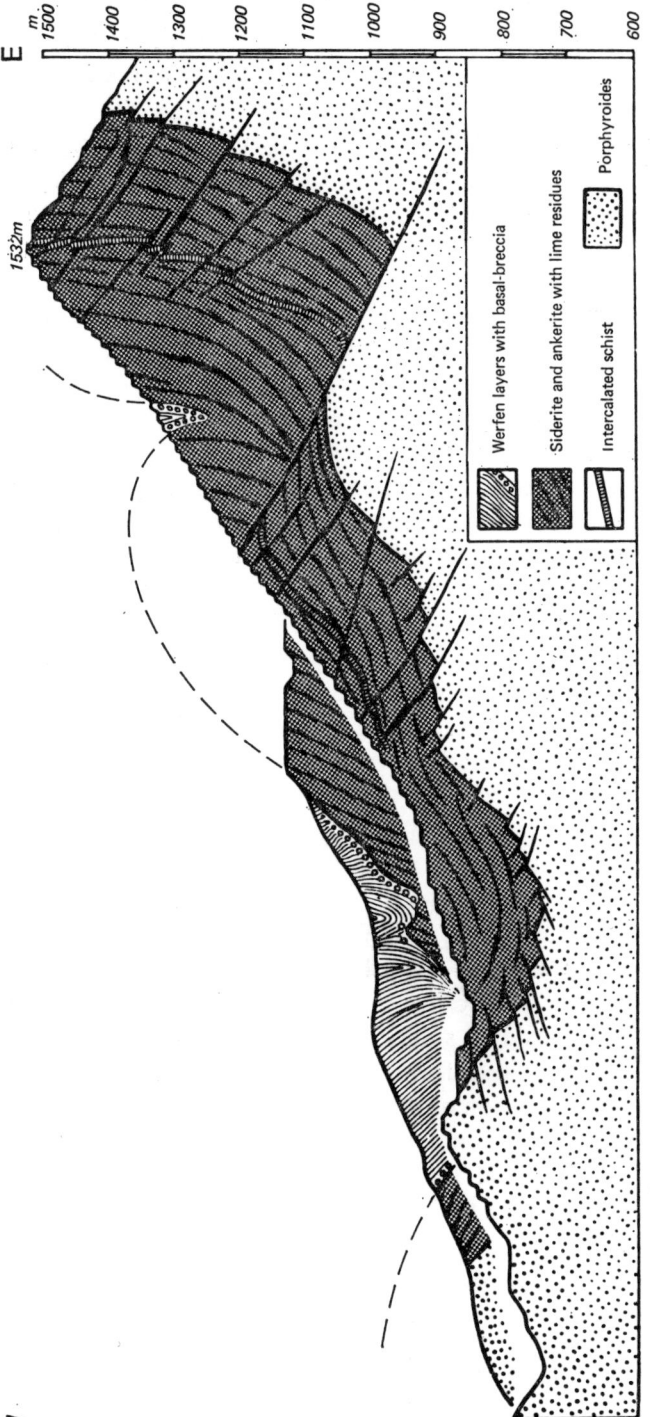

FIG. 25. Cross-section of the Styrian Ore Mountain (Austria) (from Schneiderhöhn, 1941).

transformed later to goethite (red iron hydroxide or 'campanil'). In Cumberland, England, iron-ore stocks composed of siderite and hematite occur in Carboniferous limestone.

Impregnation deposits are a further hydrothermal structural form. They will develop when the hydrothermal solutions (especially in silicate rocks of low solubility) deposit their mineral content mainly in rock pores, on grain boundaries and cracks and thus 'impregnate' the rock. The deposits thus formed, being finely disseminated, have relatively low metal contents. Impregnation deposits are usually stock-shaped, in which, of course, the location of the fissures, conducting the solutions, and the suitability of porous layers play a decisive role. The most commercially important impregnation or disseminated deposits are confined to quartz monzonites, granites and porphyritic rocks. Impregnation deposits are of interest for the miner because they often possess extraordinarily large dimensions and notwithstanding their low metal content may contain rather important amounts of metal. Metallization mostly results in iron and copper sulphides (pyrite, chalcopyrite). The copper deposits of *Bingham*, Utah, USA (porphyry copper ores), range among the largest copper-ore deposits of the world. They are chalcopyrite-pyrite impregnations in the domes of granite (monzonite) massifs (Fig. 26). The Bingham production is 225,000 tons of broken-up material per day, which corresponds to 90,000 tons of raw ore per day. Since the discovery, in 1900, the total material removed amounts to 2,500,000,000 tons. The open-cut mine is about 500 m deep; about 20% of the total copper production of the USA comes from here.

These three structural forms of the hydrothermal deposits have one feature in common—they are all intracrustal bodies, that is they are situated within the solid crust.

Hydrothermal residual solutions from a magma can be exuded also at the sea bottom and form submarine bodies. The minerals are precipitated by the action of the sea at the sea bottom and its sediments. Thus the formation of commercially important *submarine hydrothermal deposits* is also possible. These deposits formed on the sea bottom are mainly characterized by their being always located in geosynclinal areas of the earth. There they occur frequently with tuffs, lavas and intrusions of basic (to acid) igneous rocks (for example, diabases or dolerites, keratophyres, spilites, diabase-tuff or 'schalstein'), alternating with marine sediments. These are rocks that are derived, on the one hand, from submarine volcanism (endogenetic cycle) and represent, on the other, normal marine sediments (exogenetic cycle). The structural form of these deposits is controlled by the sedimentary structure at the sea

FORMATION BY MAGMATIC PROCESSES 43

bed. Thus we have ore lenses and 'seams', alternating layers of ore and shales or tuffs (Fig. 27).

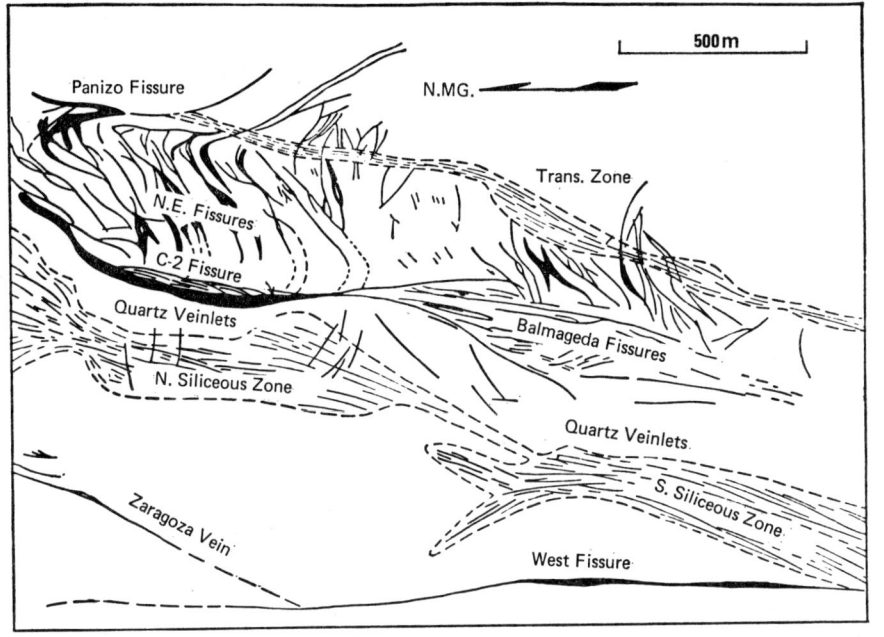

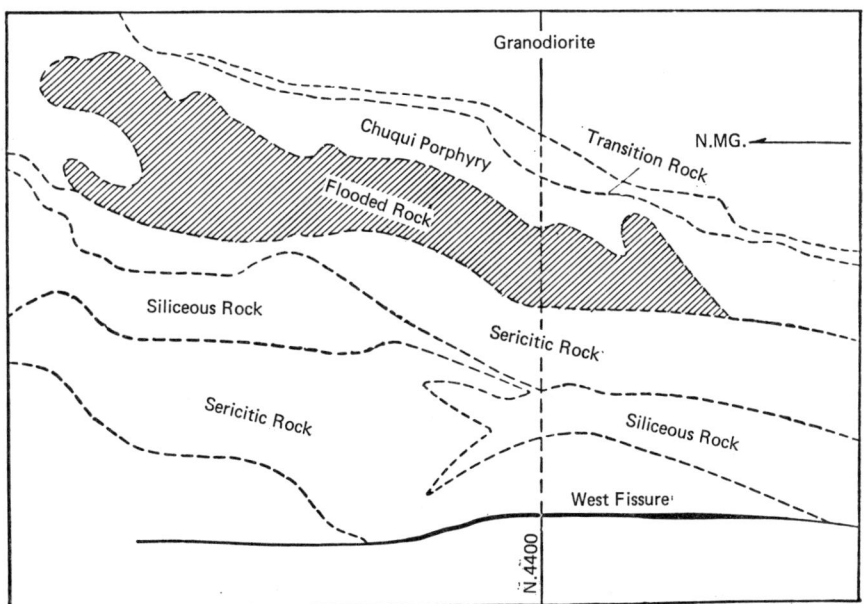

FIG. 26. Sketch map of tectonic (a) and geological (b) setting at Chuquicamata, Chile (from Bateman, 1951).

The mineral content of these submarine hydrothermal deposits may be represented by oxides or sulphides, depending on the redox potential prevailing at the sea bottom. As an example of an oxide body, the iron-ore deposits of the *Lahn-Dill type* may be mentioned. They belong to the most important iron-ore districts of West Germany. They were formed during the Middle Devonian (Variscan geosynclinal phase). The ore deposits, consisting mainly of hematite, are confined to the so-called 'schalstein'-series (tuffs and diabases with limestones). Besides this predominantly magmatic schalstein facies, there is also developed the facies of the massive or Stringocephalus (coral-reef) limestones. Some of the iron ore lies in the form of bedded layers and elongated lenses, from 2 to 6 m thick, inside the schalstein series. The main ore deposit, however, occurs in the roof of the schalstein series as a larger compact horizon. Then follow the overlying limestones of the Upper Devonian.

In an environment similar to that of the iron-ore deposits of the Lahn-Dill type metallic sulphide gels may be deposited on the sea floors, with low redox potentials (H_2S facies) by the separating out of H_2S solutions and metal-bearing solutions. There will then form submarine hydrothermal sulphide deposits containing pyrite and varying amounts of chalcopyrite, sphalerite and galena of the *Rio Tinto-Meggen type*. In this way the largest pyrite deposit in the world, that of Rio Tinto, Huelva district, Spain, originated. This vast district of ore deposits includes parts of southwestern Spain and southern Portugal. In a Silurian-Carboniferous complex of shale, trending west-east, intrusive and extrusive igneous rocks are interlayered (diabases, keratophyres and others associated with geosynclinal magmatism).

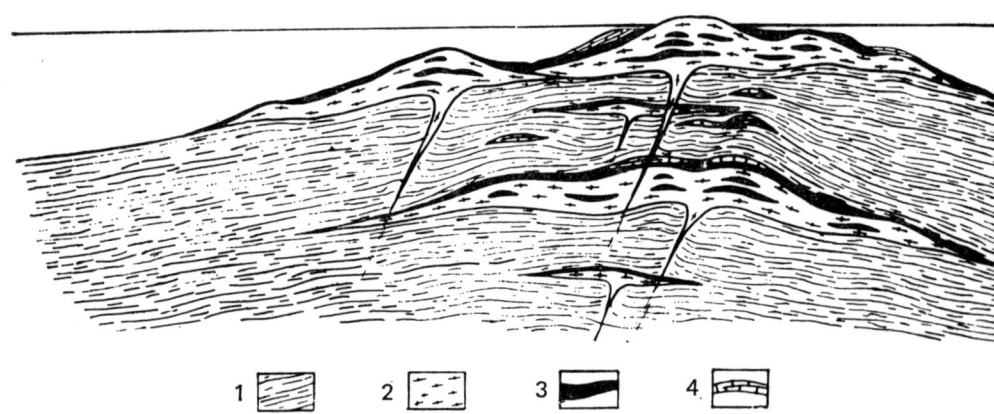

FIG. 27. Cross-section of a submarine hydrothermal sedimentary deposit (area of Sternberk-Hor. Benesov/CSSR.) (after Skacel, 1966).

1–Devonian schists,   2–volcanic series,
3–iron-ore horizons,   4–limestones.

Associated with these rocks—in an area 100 × 40 km in extent—are numerous lens-shaped, partly massive, pyrite deposits (up to 1000 m long and 150 m thick). The ore bodies are often of surprising purity (up to 90% of pyrite). Moreover, there occur smaller amounts of chalcopyrite (2-3% Cu) and several other sulphides. Due to the large amounts of pyrite with some chalcopyrite Rio Tinto is also one of the main producers of Cu in Europe. The ore is won in large open-cast mines and also by underground mining. The output is about 2-3 million tons per year. The present reserves are reported to be 500 million tons of ore (Fig. 28).

The pyrite-sphalerite-baryte deposit of Meggen, W. Germany, is also conformably intercalated with shales and limestones, in this case of Middle Devonian age. The ore deposit is 3 m thick and originally had the shape of an elliptical lens in synclinal form. It is remarkable that the pyrite deposit, towards its original edge, gradually changes into a baryte deposit. The metallization is derived from a magma body located at a depth of about 1000 m, and the hydrothermal liquids reached the basin of deposition by way of marginal tensional fissures (Fig. 29). The Meggen area was subjected to strong deformation during the Variscan orogenesis and divided into many blocks. Therefore the whole ore deposit today forms a more or less imbricated double syncline.

Genetically, parts of the metamorphic skarn iron-ore deposits of central and northern Sweden also belong to this group. The ore deposits frequently occur together with lime silicate rocks within the Precambrian 'leptite formation' (acid lavas and tuffs of geosynclinal origin). The most important district is that of *Kiruna* in northern Sweden. The ore is predominantly a fine-grained to dense magnetite very closely intergrown with apatite. Hematite is very much reduced in amount. The iron content varies from 50 to 70% (P content up to 5%). In the topmost parts of the deposit intercalation of magnetite with apatite takes place; banded hematite-quartz portions also occur (iron jaspilites, e.g. the 'Rectoren' deposit). The enormous iron-ore deposit of Kiruna has an average thickness of 70 m, an extension along the strike of 15 km and a proved depth down to 1000 m (dip 50-60° east). With reserves amounting to 2000 million tons Kiruna represents perhaps the largest accumulation of rich iron ores in the world. Morphologically the ore body forms a mountain ridge, which has become a 'split mountain' due to open-cast mining operations (Fig. 30). Further magnetite-hematite-apatite deposits of this kind within the leptite formation are those of *Gaellivare*, North Sweden, and *Graengesberg*, Central Sweden.

46 INTRODUCTION TO ORE DEPOSITS

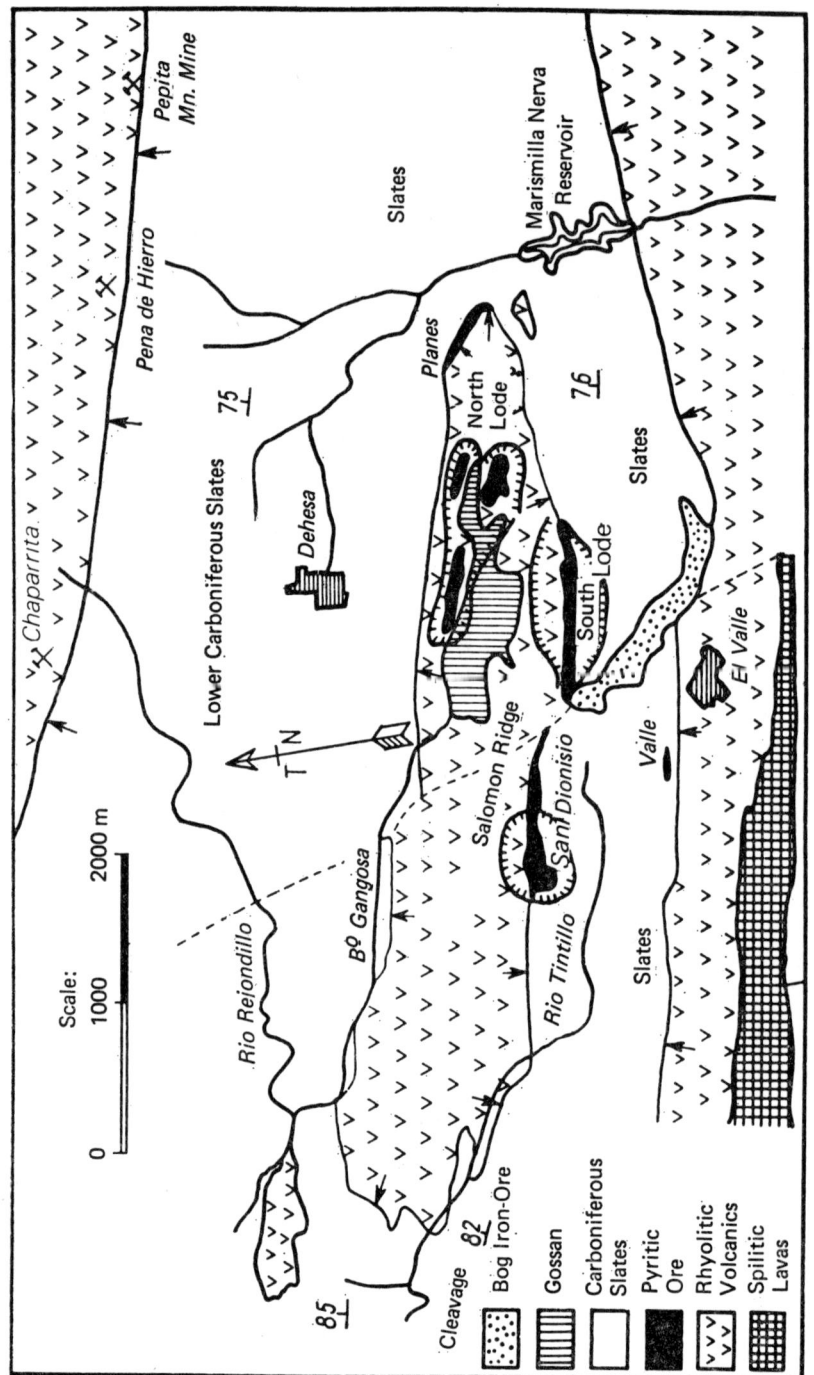

FIG. 28 a. Geological plan of Rio Tinto area, Spain (after Williams, 1965).

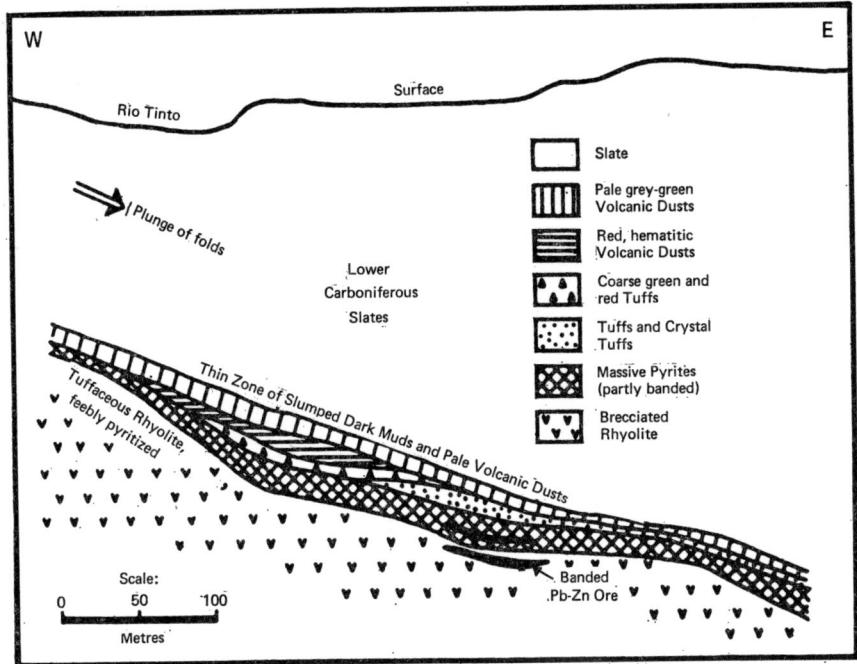

Fig. 28 b. Diagrammatic section along axis of New Planes Orebody, Rio Tinto (after Williams, 1965).

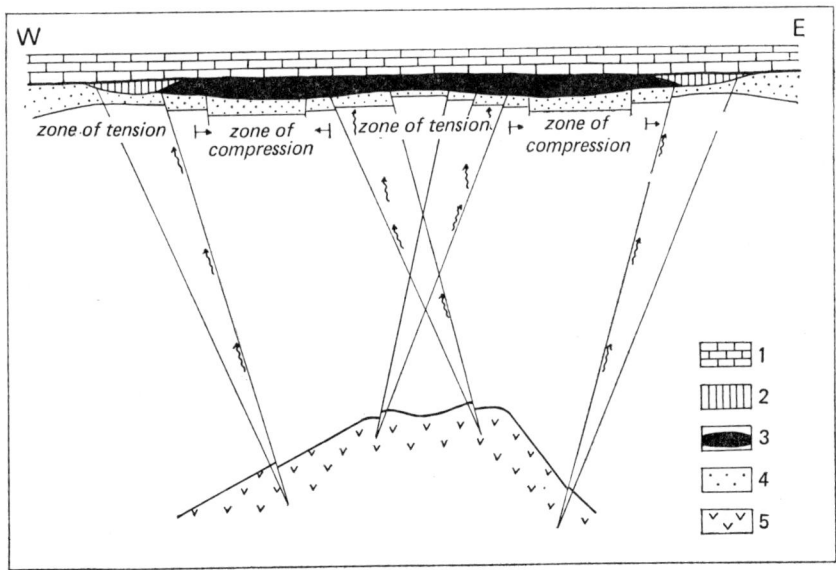

Fig. 29. Schematic section of pyrite-sphalerite-baryte deposit of Meggen, West Germany.

1–Upper wall of limestones (Middle Devonian)
2–Baryte body
3–Ore body (FeS_2, ZnS)
4–Lenne schist with intercalations of limestone and pyrite
5[1]–Intrusive body.

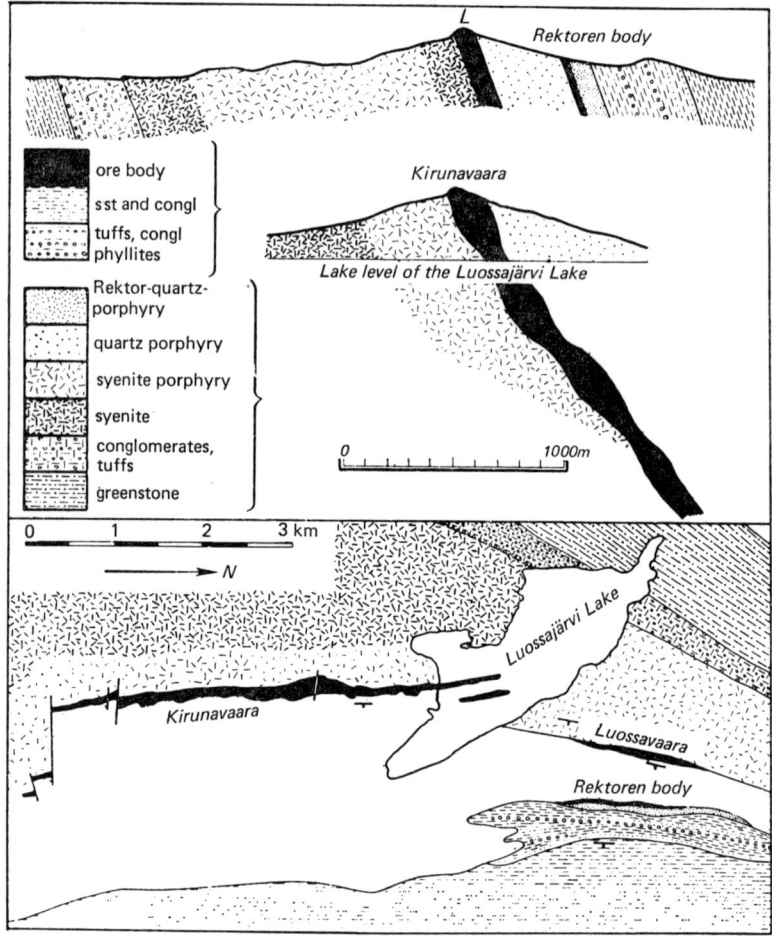

FIG. 30. Iron ore district of Kiruna, Sweden (from Schneiderhöhn, 1958).

1.4 Spatial distribution of magmatic deposits

1.4.1. Zonal distribution

Experience has shown that ore deposits undergo changes in mineral composition with increasing depth. These changes are due to varying physico-chemical conditions as the magmatic source is approached. Normally, the spatial arrangement of mineralization corresponds to the succession of mineral precipitation. A simple transition is found from near-source, older, high-temperature minerals to distal deposits of low-temperature (epithermal) origin.

Changes in facies or zoning occur both vertically and horizontally (Fig. 31a) and are referred to collectively as *lateral facies changes*. The vertical change, which is especially common, has also been called the 'primary downward change'. If the individual mineral assemblages are bedded, not adjacent to one another, but superimposed upon each other, then they represent a *'temporal facies change'* (Fig. 31b). A blurring of the zonal arrangement can occur on quick cooling of the magma. When this happens mineral sequences formed at lower temperatures affect an interior area composed of early, high-temperature mineral-assemblages. This superimposition of zones is referred to as 'telescoping'.

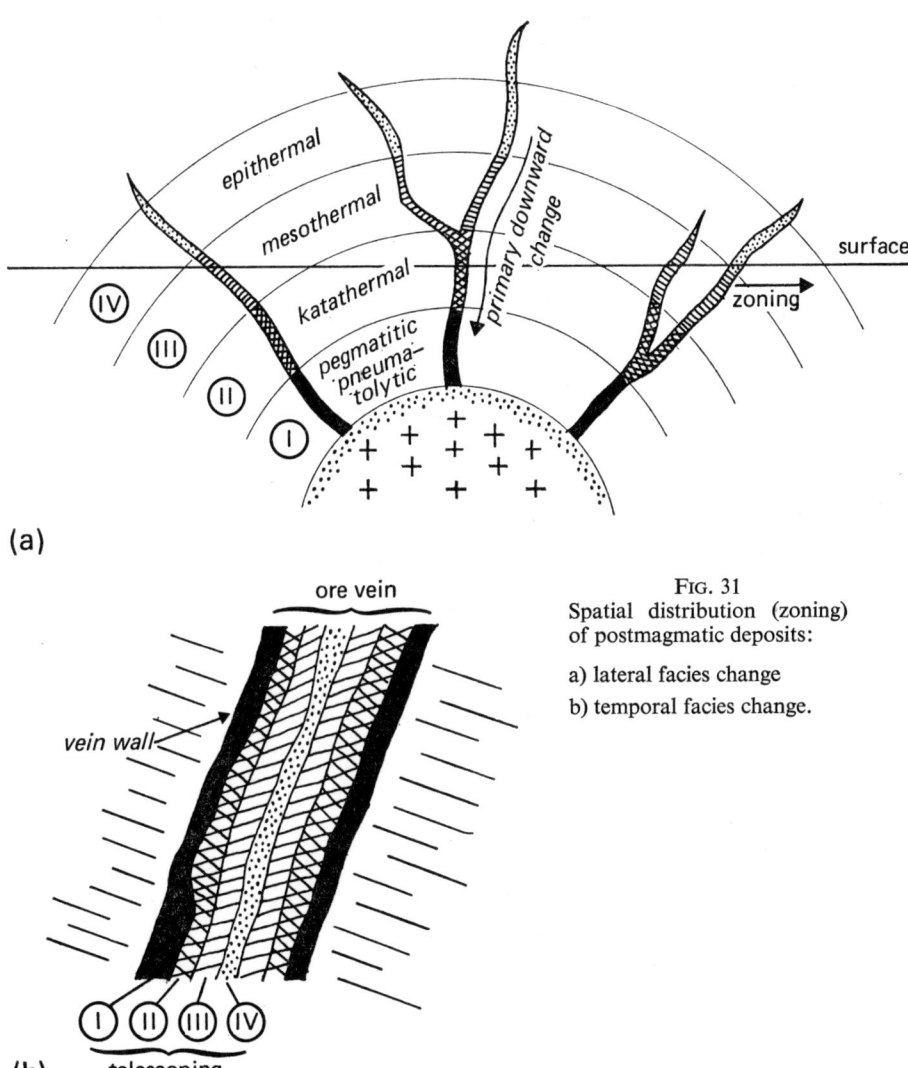

FIG. 31
Spatial distribution (zoning) of postmagmatic deposits:

a) lateral facies change
b) temporal facies change.

Examples of such a distribution of minerals (zoning) with stockwork- or aureole-like arrangement around the magmatic source pluton are found in nearly all districts (Fig. 32).

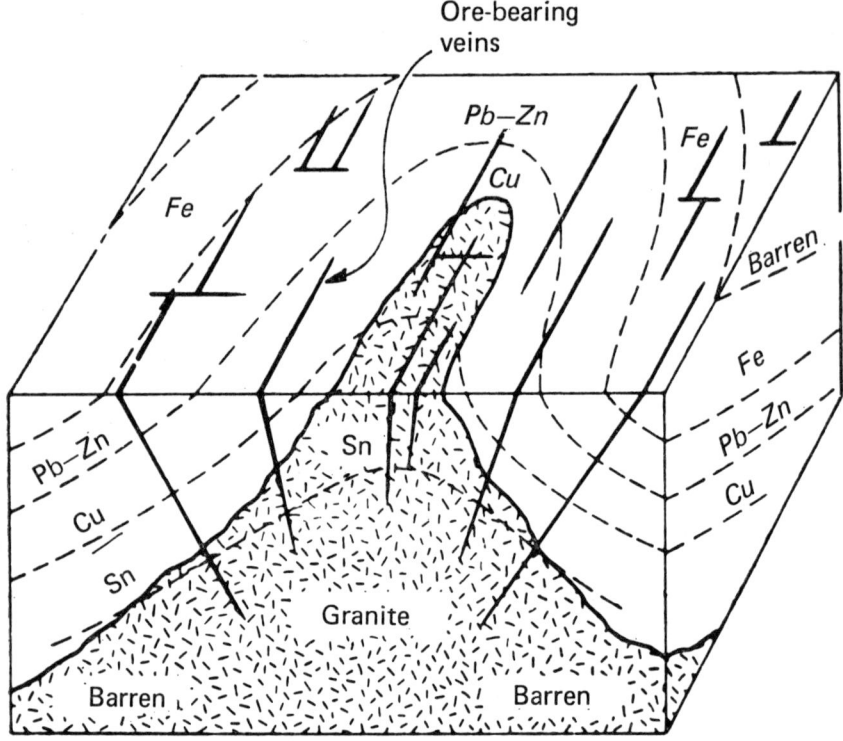

FIG. 32. Block diagram showing the arrangement of ore zones relative to granite-slate contact, Cornwall, England (from Park and MacDiarmid, 1964).

Other factors controlling the nature of the ore deposits are

(a) Level of Intrusion;
(b) Area of Precipitation;
(c) Tectonics.

1.4.2. Level of intrusion

We have commented on the control which pressure and temperature have on mineral formation. Since both factors increase with depth then the level of intrusion of the magma is important (Fig. 33). We can distinguish *plutonic*, *sub-volcanic* and *volcanic* (or exhalation) ore deposits. The depth of formation of plutonic deposits varies approximately between 3000 and 500 m below the earth's surface. The subvolcanic deposits are formed approximately between 1000 and 100 m. As the temperature of consolidation of a granitic pluton is about 700 °C and the temperature at the earth's surface is between 10 and 20 °C, the temperature gradient over a deep-seated pluton is relatively small; with a high-level subvolcanic area, however, a very steep temperature gradient occurs. Accordingly, the mineral parageneses in the slowly cooling deeper-seated plutonic levels are mostly distinctly separated from each other locally and are very distant from each other in the vertical sense. In the high plutonic and the subvolcanic levels cooling takes place relatively quickly due to the proximity of the surface, and the mineral parageneses show strong 'telescoping'.

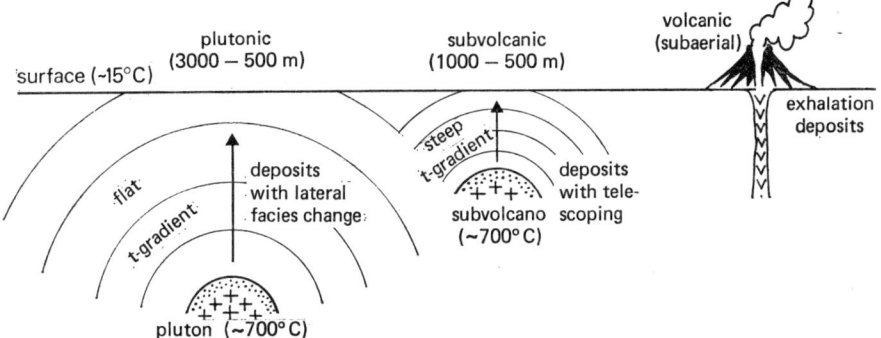

FIG. 33. Variation in character of ore-deposits according to the level of intrusion.

1.4.3. Area of precipitation

Besides different levels of intrusion we can also distinguish different places of precipitation or accumulation. The latter may be *intramagmatic*. Here the formation of deposits takes place within the magma body itself. These are liquid-magmatic deposits which originate in the course of crystallization differentiation or the separation of liquids because of immiscibility (see pp. 13–19).

Following residual-liquid differentiation, that is at the pegmatitic-pneumatolytic and hydrothermal stages, the formation of *intracrustal* deposits may take place. These are chiefly associated with the topmost zone or the overburden of the intrusive body (in vein fissures, in the form of impregnations or as metasomatic replacement bodies). In the formation of these 'epigenetic' deposits the influence of the country rock may sometimes become so considerable that the character of the ore deposit is thoroughly changed. Such deposits are designated as 'topomineral reaction deposits' and originate by the reaction of the ore solution with the country rock.

With exhalations in marine areas of the metal-carrying residual liquids *submarine* deposits will be formed. These deposits, which are assumed to be syngenetic with marine sedimentation, are mostly connected with submarine volcanism. In these 'mixed' hydrothermal-sedimentary deposits the metal content is derived from endogenetic processes, whereas precipitation and deposition are determined by exogenetic processes. They appear in the form of layers, in which beds of lavas alternate with those of tuffs and marine sediments. As regards structure on a large scale the ore-bodies tend to be stratigraphically restricted but on a small scale they are variable. They may be lenses wedging out relatively quickly or horizons of deposition of considerable regional extension. Numerous very important deposits of iron ores and nonferrous metals belong to this group (for instance, the types of Lahn-Dill and Rio Tinto-Meggen). During the deposition of the volcano-genetic metals on the sea bottom, the ore solutions may have an effect on the syngenetically formed marine sediments. There occurs, in addition to the syn-sedimentary ores, therefore, local forms of metasomatic replacement.

In the case of *subaerial* emanations of the metal-bearing residual liquids, that is of volcanic origin, the elements escape into the atmosphere as vapours, and are rarely concentrated into ore bodies.

1.4.4. Tectonics

(*a*) *Regional tectonic pattern.* As in magmatism the ore deposits are influenced in their spatial distribution also by the large-scale tectonic pattern. There appears especially a connection with large anticlines as well as with major faults. These are the large-scale structures, which determined the path of ore-carrying magma. As an example of the accumulation in large-scale anticlines, the Alpine gold-ore veins in the nappe culmination of the eastern Hohe Tauern may be cited. A distinct example of the association of ore deposits with large-scale fault lines is provided by a number of large deposits of the Colorado Plateau, USA (Clifton Morenci, Miami, Climax, Bingham, Leadville and others). These deposits are associated with deep-reaching fractures, which form the boundary of the stable block of the craton against the folded zone of the Rocky Mountains (Fig. 34).

It should further be mentioned that the intracrustal deposits of pegmatitic-pneumatolytic origin, and partly also those of hydrothermal origin, are considerably influenced by the mechanical tectonic processes involved in the emplacement of plutons. The mechanism of solidification that is especially important in the 'emplacement' of ore deposits for example in the formations of 'stockscheiders', joints and veins in the granite (Fig. 35).

In the overlying zones of a pluton it can often be found that the different phases of mineralization separate out in vein fissures located in different regions and having different directions of strike. This means that the tensile stresses in the top of the pluton changed as regards space and direction during the precipitation of the residual liquids. This is mainly caused by the irregular granite morphology. Because of this, the loss of heat and, simultaneously, the rate of solidification of the magma body are different in different places. Therefore, the centre of the reduction in volume does not only change vertically, but also horizontally (Fig. 36). The top zone is therefore influenced during the solidification of the granite by differently directed tension vectors that open differently directed joints to form vein fissures. The centre of

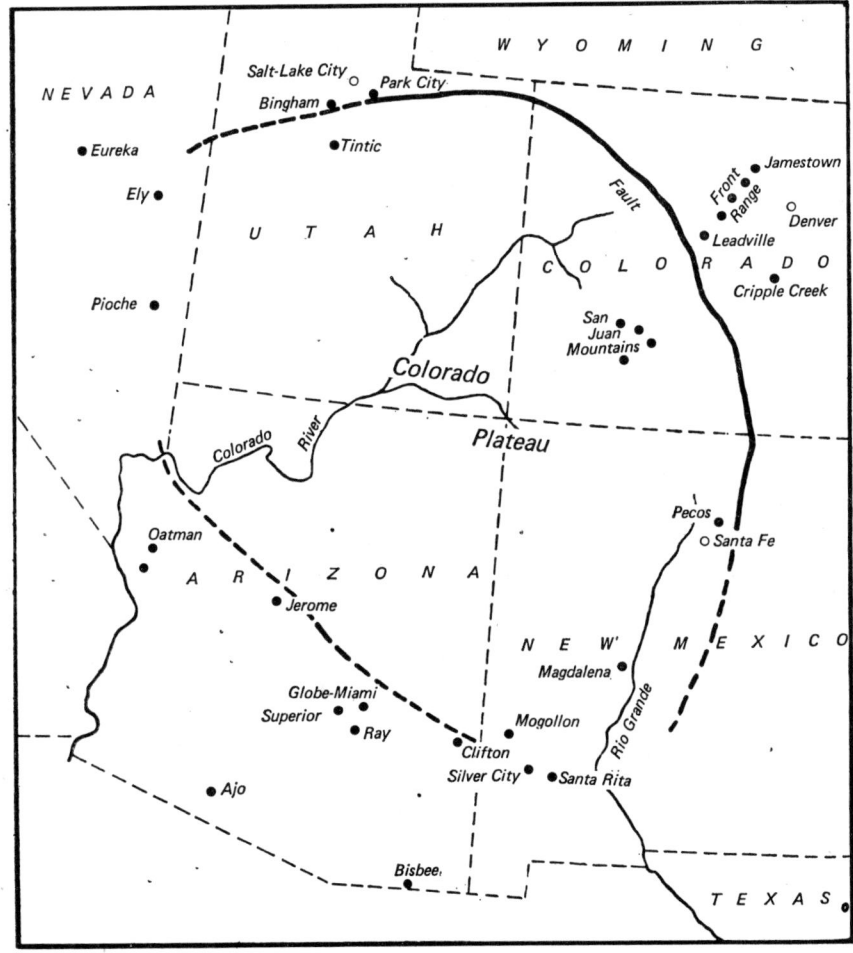

Fig. 34. Deposits association with the Colorado Plateau fault zone (from Petrascheck, 1961).

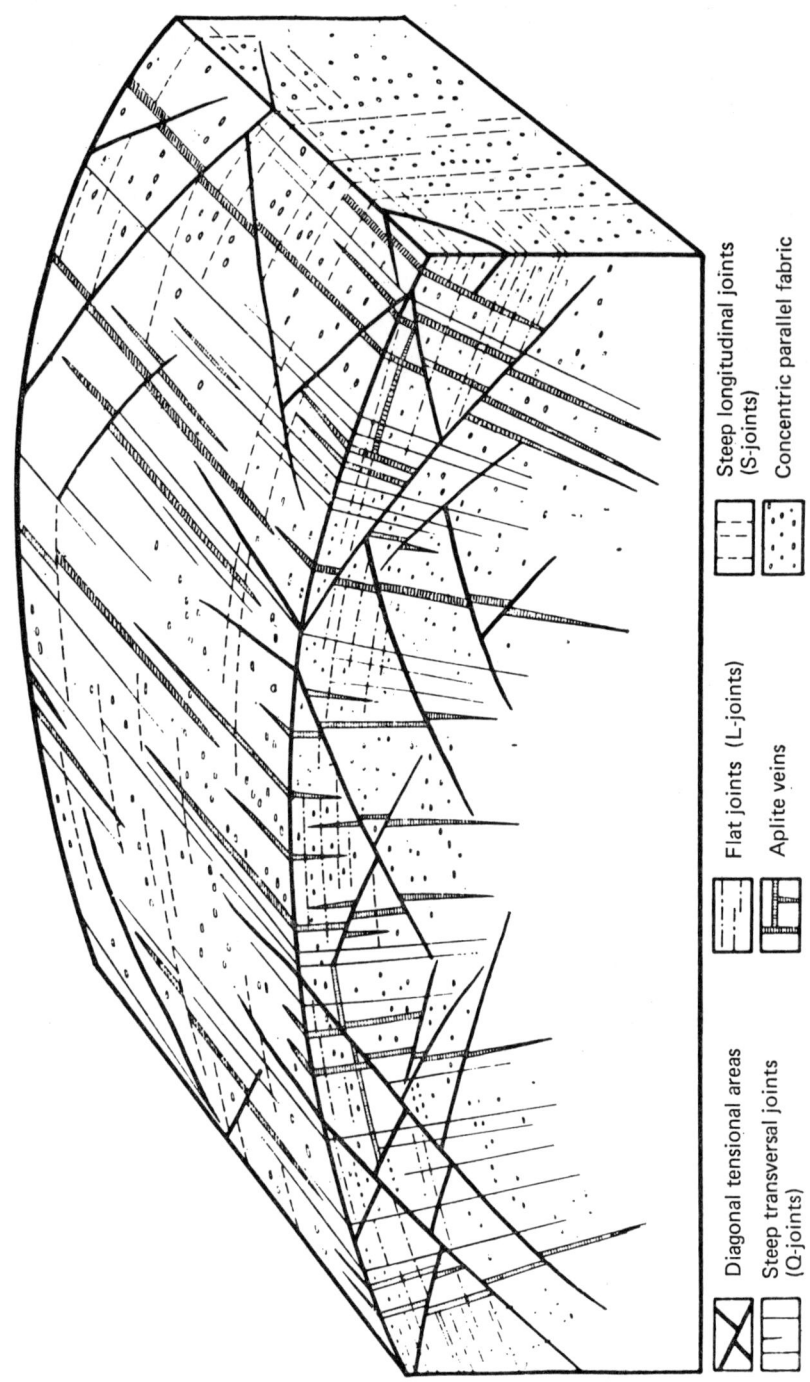

Fig. 35. Regular intrusive body with the different phenomena related to emplacement (from H. Cloos, 1936).

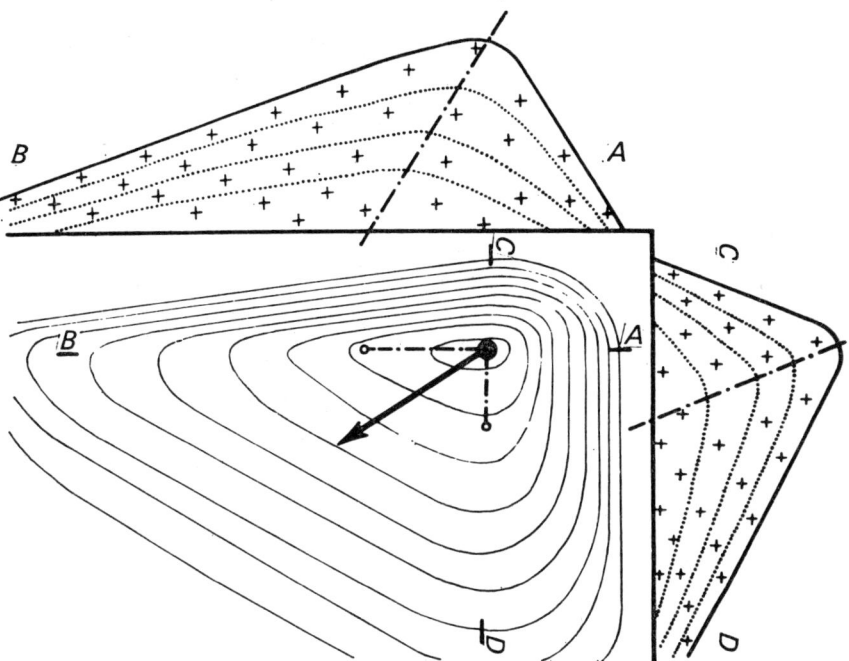

Fig. 36. Displacement of the centre of volume-decrease diagonally within flat flank of an intrusion (after Oelsner, 1958). *AB, CD* are vertical sections whose positions are indicated on the plan.

shrinkage tends to move towards the flatter margin of the intrusion. Greater subsidence occurs here, and there follows a greater concentration of residual liquids. This, in turn, means that fissures and ore veins are always richer near the flat parts and poor near the steep boundaries of a pluton.

(*b*) *Local tectonic pattern.* Evidence for the distribution of the ore in the veins (ore shoots), or the spatial arrangement of the ore bodies, is of great importance for geological exploration and the development and excavation of the deposits. Ore distribution in relatively small areas is mainly controlled by the local tectonic pattern and the kind of country rock present. It is one of the chief tasks of the mining geologist and the miner to find out these local features for each mining area.

In many ore districts there is a distinct dependence of metallization upon directions of strike or dip (Fig. 37). This phenomenon can be observed in all important ore-vein districts of the earth (Freiberg, Butte, Coeur d'Alene). In *Laurion*, Greece, and *Keban*, Turkey, the Pb-Zn metallization follows older eruptive-rock veins and then spreads metasomatically in limestone below sealing shale horizons.

The metasomatic bodies of *Trieben*, Austria, and the metasomatic siderite bodies of *Huettenberg*, Austria, have the same general trend as the fold or B-axes of the folded rocks (Fig. 38).

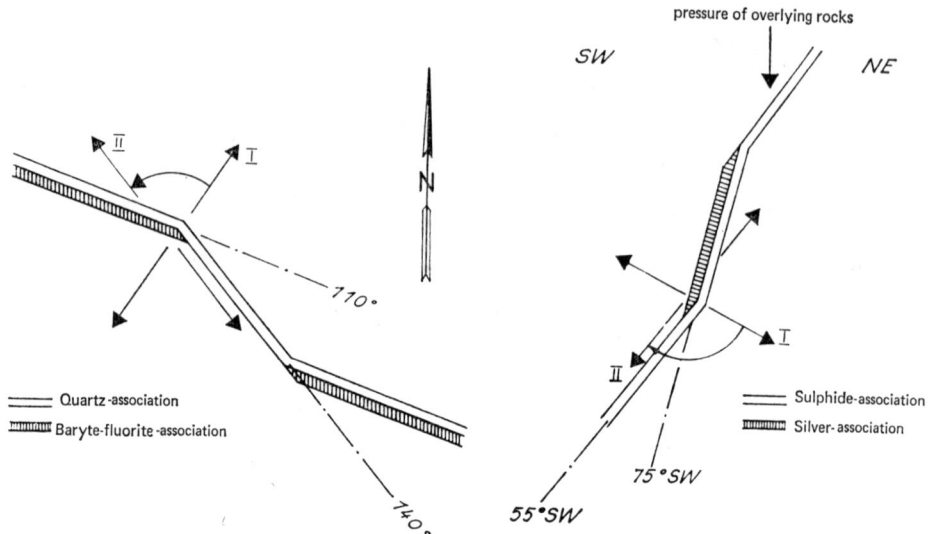

FIG. 37. Formation of ore shoots in veins (from Baumann, 1958). I and II represent the sequential and directional opening of the ore veins.

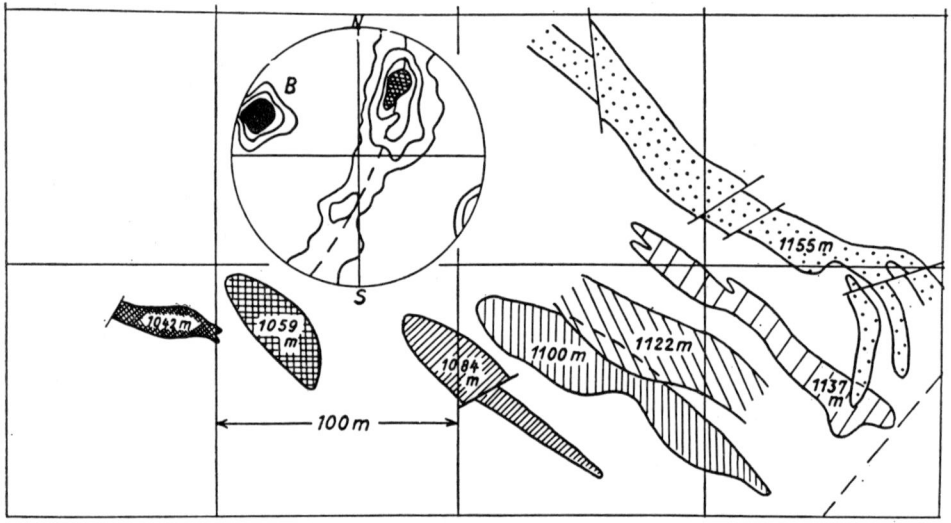

FIG. 38. Metasomatic bodies of Huettenberg oriented parallel to the fold axes (*B*) (from Metz, 1967). The inset stereogram shows the position of *B* and the great-circle distribution of bedding planes. The ore-bodies represented lie at different levels and the elevations are given above sea-level.

1.5. Ore deposits, tectonics and magmatic activity

1.5.1. The geotectonic-geomagmatic cycle

Several large-scale cycles, based on tectonism and magmatism, have been recognized in the course of the earth's history. Magmatic activity and tectonic development are interlinked and according to Stille (1940) the connection is as follows:

Major structure	Magmatism	Origin of magma
Geosyncline	initial	simatic
Orogen	synorogenic / late-orogenic / post-orogenic	sialic
Kraton	final	simatic

Geosynclines are large, elongated areas of subsidence bordered either symmetrically or asymmetrically by stable continental (cratonic) masses. The asymmetric type lies on the margin of a continent with a cratonic mass on one side and an ocean basin on the other. Some sediment is supplied to the geosyncline from the craton but even more derives from islands (cordilleras) which arise within the trough. Subsidence and sedimentation tend to be continuous. Estimates of the accumulation of sediment vary but it may be about 25 m per million years. Ore deposits may be associated with these sediments. These are described in detail below.

Geosynclinal sedimentation is supplemented by the intrusion of basic magmas. These magmas occur in the form of gigantic, flat intrusive bodies and bedded veins as well as submarine effusions. With this *initial magmatism*, juvenile material is involved rising along enormous zones of tension from great depths (Fig. 39). Thus, the material consists of products of the differentiation of basaltic magmas. These events are followed by the formation of manifold types of deposits of initial magmatism (intramagmatic, intracrustal, submarine).

The geosyncline represents a mobile zone of weakness, which is, finally, overcome by the tangential pressure of the adjacent stable continental masses. This marks the beginning of orogeny, short geologically but violent, with its phenomenon of intensive folding (for example, the Alps between Europe and Africa, the Urals between Europe and Asia). The *synorogenic magmatism* is characteristic of the first part of the orogeny. The material is mostly of palingenetic origin from the Upper Crust. It is generated by remelting (i.e. palingenesis), when sial is buried at depths of 20-25 km and thus mobilized. The upward intrusive force of these molten masses, whose amount is

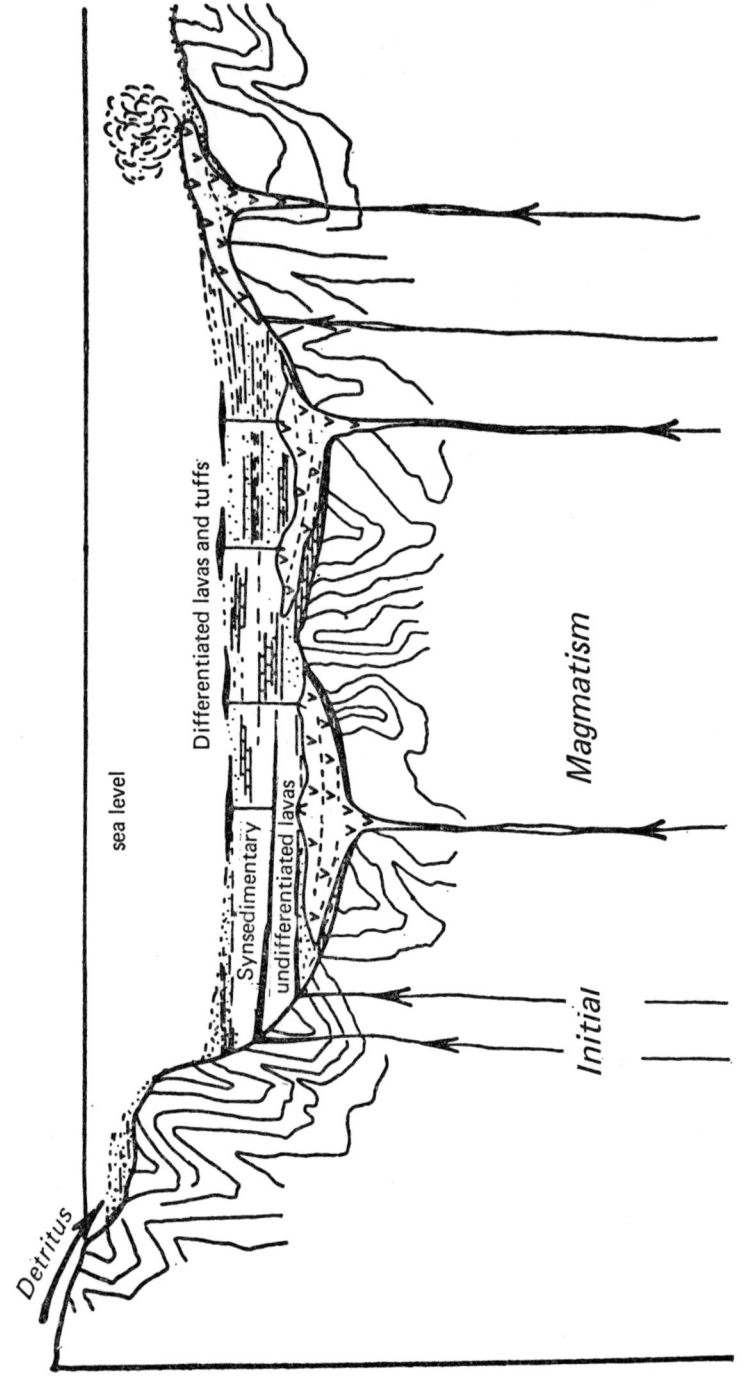

Fig. 39. Schematic section of a geosynclinal part of the earth's crust (after Borchert, 1960).

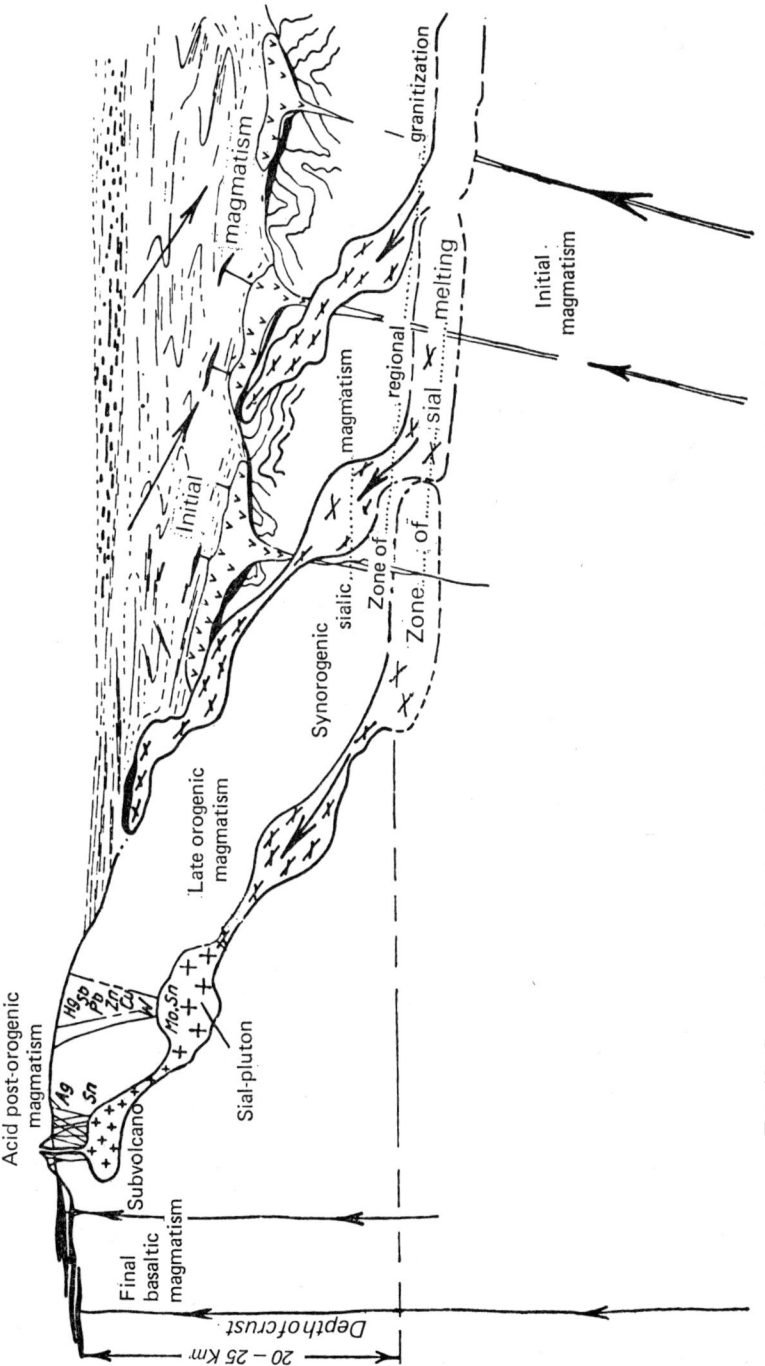

FIG. 40. Succession of simatic and sialic magmatism in time and space (after Borchert, 1960).

proportional to the depth of the geosyncline, is caused by the difference in density between the solid rocks (2·70-2·85) and the sialic melt (2·55). Increasing consolidation and elevation of the formerly mobile geosynclinal zone take place both by folding and by intrusion. At this time *late-orogenic* magmatism processes of concentration are developed. These will form ore deposits of *plutonic origin* revealing classic ore zonation (Fig. 40). The following *post-orogenic magmatism* is characteristic of the beginning of the cratonal state when the previously mobile area stabilizes. Then precipitation of ore deposits of *subvolcanic origin* with more marked telescoping takes place (Fig. 40).

After the final consolidation of the *kraton* the *final magmatism* becomes characteristic. The material of the final magmatism is juvenile basaltic magma again, derived through deep-reaching fracture zones. Some of the basaltic magma may congeal within the middle zone of the earth's crust. It may be subjected to fractional crystallization and give off residual liquids on slowly cooling (Fig. 40).

Due to later, slow subsidence of the kraton its deeper regions may reach the zone of *metamorphism* where the deposits contained in it become re-deposited metamorphically or mobilized.

Based upon these two geomagmatic cycles (simatic-basaltic magmatism of juvenile origin and sialic-granitic magmatism of palingenetic origin) the magmatic ore deposits can be genetically divided into:

(*a*) Deposits of simatic magmatism;
(*b*) deposits of sialic magmatism.

1.5.2. The types of ore deposits of simatic magmatism

The most important types of ore deposits, associated with juvenile basaltic magmas and their products of differentiation, are as follows. The deposits due to primary crystallization are best developed. Here we have gravitative precipitation of *intramagmatic* deposits of chromite, nickeliferous pyrrhotite-chalcopyrite, platinum and titaniferous magnetite-ilmenite (=deposits of *initial magmatism* in a strict sense). In the course of residual crystallization, more intensive enrichment of volatile constituents takes place at the top of the basic intrusive bodies. We might refer to these as deposits of a '*prolonged*' *initial magmatism*. The metal-bearing residual liquids rise upward and may be extruded into the geosynclinal areas, mainly at the sea bottom. In accordance with the geotectonic pattern, initial magmatism is followed by a preferential precipitation of *submarine* deposits (iron-ore deposits of the Lahn-Dill type, iron silicate deposits of the chamosite type, manganese-ore deposits of the radiolarite type, sulphide deposits of the Rio-Tinto-

Meggen type). The metal content of these deposits is mainly confined to the elements Fe, Mn, Cu, Zn, Pb, and Ba, Ni, Co.

These deposits may frequently be transformed by subsequent regional-metamorphic episodes. For example, the Lahn-Dill type may be transformed into the magnetite deposits of the Kiruna type. The Rio-Tinto-Meggen association on the other hand may change to metamorphic pyritic like Leksdal, Norway; Röros, Norway; and Falun, Sweden deposits. Besides the submarine deposits there are corresponding *intracrustal* deposits. The farther the residual liquids (at the top of the intrusive body) are from the sea bottom or the ground surface, the more chances there are for the ore and accessory minerals to be precipitated in previously formed geosynclinal sediments. Thus we have formed vein deposits, impregnation deposits, porphyry copper-ores and metasomatic replacement deposits. Here, too, the element content is principally confined to Fe, Mn, Cu, Zn, Pb, Ni, Co and Ba. The deposits form the transition to the ores of the *final magmatism* of those parts of the crust which have become kratonal. The mineral content of these deposits reveals largely the same combinations of elements as the deposits of the initial magmatism (Ba, F, Fe, Mn, Cu, Co, Ni). Most baryte-fluorite veins as well as the iron-manganese-ore veins belong to them. Furthermore, a great part of the vein deposits of the Bi-Co-Ni-Ag formation is also associated with the final basic magmatism (e.g. Erzgebirge, Black Forest, Cobalt City).

This general outline shows that many ore deposits have been derived from juvenile simatic magmatism; in fact it has played a predominant part in ore deposits formed by magmatism.

1.5.3. The types of ore deposits caused by sialic magmatism

The sialic magmas are mainly remelted sedimentary and other materials of the upper crustal layers, including the metallic components already present there. During the crystallization of these palingenetic melts, acid granitic rocks mainly are produced together with extremely acid silicate-carrying, and partly also metal-bearing, residual liquids at the stage of residual crystallization. For reasons already explained considerable intramagmatic metal concentrations do not form during the primary and main crystallizations. During residual crystallization, however, numerous very rich metal deposits may be generated. For geotectonic reasons submarine precipitation scarcely plays a role in this phase, the geosynclinal phase having passed. Here appear the most classic and important *intracrustal* deposits of the supercritical phase and partly also those of hydrothermal origin. First of all the pegmatitic deposits comprising of feldspar, quartz and mica pegmatites as well as pegmatites with precious stones and rare earths. Then the pneumatolytic Sn-W-Mo deposits (with tourmaline) and the whole series of the metal-rich hydrothermal deposits belong here (starting with the Au-Ag formation, through the pyrite-tourmaline-chalcopyrite formation, Pb-Zn-Ag formation, U-Fe formation, to the Sb-Hg formation). The mineralization may occur both in plutonic situations and in subvolcanic structures.

The kind of metallization brought about by the sialic-palingenetic magmatism depends principally on the material subjected to melting under the geosyncline in the course of the orogenic stage. Due to the sedimentary source material being very various, the ore-deposit content is much more diversified than that resulting from simatic magmatism.

2
Formation of ore deposits by sedimentary processes (exogenetic cycle)

2.1. The importance of the hydrosphere, atmosphere and biosphere

The formation of ore deposits of the endogenetic cycle requires a magmatic melt with its processes of differentiation. The development of the exogenetic or sedimentogenic deposits on the other hand takes place mainly under the action of the *hydrosphere, atmosphere* and *biosphere* with their manifold controls of concentration and selection. The processes of the exogenetic material cycle promote the adjustment of the outer layers of the *lithosphere*, that is the upper parts of the earth's crust, on the one hand, and the spheres operating at the earth's surface, on the other (Fig. 41). It goes without saying, that the rock constituents of the lithosphere, produced under quite different conditions of formation (magmatic and metamorphic-diagenetic conditions), cannot remain stable at the earth's surface with its specific conditions but tend to adapt themselves to these changed conditions.

The *active factors* within the hydrosphere are the water and the salts dissolved in it; within the atmosphere they are chiefly oxygen and carbon dioxide. Within the biosphere they are represented by the lower and higher organisms of plants and animals with their products of metabolism. The most active factor is *water* with its abnormal physico-chemical properties. Its great dissolving force is mainly caused by its high dielectric constant. A great role is also played by the pH-value and the redox potential (Eh-value). In such an environment and at a pressure of 1 atm. and an annual average temperature of 15 °C all silicates, oxides, carbonates and sulphides, formed by the endogenetic cycle, become more or less unstable; that is, they are dissolved at different

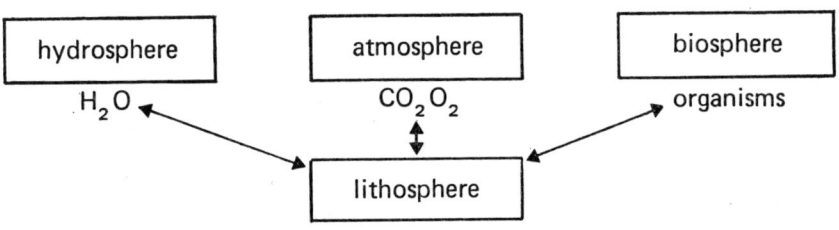

FIG. 41. Outline of the exogenic cycle.

rates, either completely or only in part. The newly formed solutions and their residues are separated from each other and partly transported. By re-deposition and re-precipitation, completely different mineral assemblages are formed which tend to be stable under conditions obtaining at the surface of the earth.

According to these processes we can divide the *exogenetic 'metabolism'* into two stages:

1. weathering (breakdown and part transformation of the minerals) and
2. sedimentation (transport and re-precipitation of the dissolved constituents).

The two stages, of course, are intimately interconnected with one another.

2.2. Fractionation and deposition

Analogous to the processes of magmatic differentiation in the endogenetic cycle we have also differentiating processes under conditions at the earth's surface; these may result in definite material concentrations and thus in the formation of sedimentary ore deposits.

2.2.1. Weathering and fractionation of the elements

The term weathering includes the changes of the rocks and minerals under the conditions prevailing at the *surface of the lithosphere*. These conditions, although variable in detail, remain as a whole within narrow ranges of pressure and temperature (1 atm. $-30°$ to $+60$ °C). Surface weathering may be classified (by the environment) as follows:

(*a*) Subaerial weathering (with access of air; main contact: lithosphere—atmosphere);
(*b*) Submarine weathering (under a cover of water; main contact: lithosphere—hydrosphere) also referred to as halmyrolysis.

The decisive factors of weathering are physical on the one hand and chemical (including biochemical) on the other.

2.2.1.1. *Physical weathering*

Physical weathering results in a purely mechanical comminution of rocks without change in chemical composition. Mechanical disintegration is largely caused by climatic factors. For example, alternate freezing and thawing of water in fissures within rocks with the accompanying changes in volume tends to break up large masses of rock in colder areas. The development of salt crystals in fissures in warmer climates has a similar effect in forcing blocks of rock apart. The effect of rapid temperature changes is not entirely clear though many workers have claimed that the differential expansion of minerals is sufficient to break up rocks. Physical weathering is of the greatest significance for chemical weathering (which acts simultaneously or immediately afterwards) because it breaks up the rocks and thus facilitates the chemical attack.

2.2.1.2. *Chemical weathering*

The chief agent of chemical weathering is water. It may operate alone. The action of water, however, is mostly intensified by 'aggressive' constituents dissolved in it (for instance oxygen, carbon dioxide, suphurous acid and others). The chemical decay of rocks and minerals caused by weathering expresses itself in the differential uptake of elements (atoms or ions) by the water. In this way substances that had been joined together by the high temperatures and pressures of the magmatic (or metamorphic) phase are again separated to form complicated assemblages. Due to the different solubility of the elements completely new compounds result.

For these processes of weathering a *series of solubility* of the different elements has been established, which depends on their 'Ionic potential' (i.e. quotient of valency and ionic radius):

$$\text{ionic potential } i = \frac{\text{ionic charge } Z}{\text{ionic radius } r}$$

This quotient determines the acid-forming and base-forming capacities of the ions (basic properties decrease with decreasing ionic radius).

In the process of weathering those elements with i less than 3 (alkalis, Ca, Ba, Sr, Fe^{II}, Mn^{II}, Cu, Zn) form ionic solutions and are relatively independent of the pH-value.

The elements with i between 3 and 10 (e.g. Mg, Al, Fe^{III}, Si, Mn^{IV}) are less soluble. Their solubility is considerably influenced by the pH-value (Fig. 42).

Those elements with i above 10 (e.g. C, P, N, S, As) form, with the hydroxyl ions of the water, soluble anion complexes. As acids (carbonic acid, phosphorus acid, nitrous acid etc.) they exert a strong solvent action on all minerals, on the one hand, and bring about precipitation by the formation of carbonates, phosphates etc., on the other.

It is first of all differences in solubility that cause a separation of the elements. Summarizing, it can be stated for the most important rock-forming elements, that the alkali metals *Na and K* are most easily dissolved and tend to remain in solution (as soluble hydroxides, capable of migration). The solution of the elements *Si, Al, Fe and Mn*, which come next in importance and are quantitatively dominant in the earth's crust, is more difficult and is pH-dependent. The alkaline earth metals

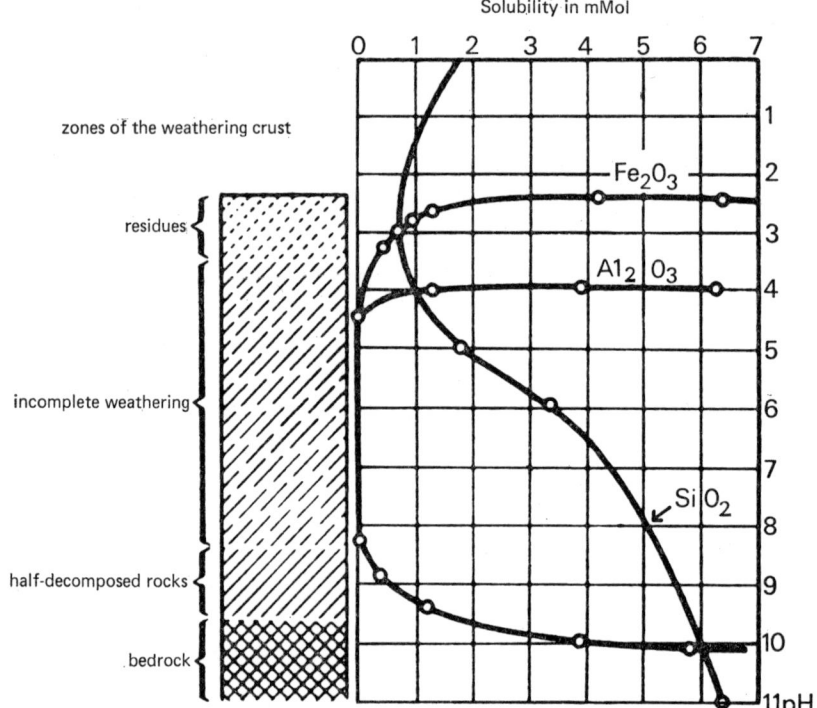

FIG. 42. Solubility of Fe^{III}, Si and Al in relation to pH-value of solution in the weathering crust (from V. I. Smirnov, 1970).

Mg, Ca, Ba, Sr lie, as regards their solubility, between Al, Fe, Mn, on the one hand, and the alkalis, on the other. They are easily soluble as carbonates or sulphates but of low solubility as hydroxides.

It has already been mentioned that the *climate* plays an important part in weathering, since it is the climate that creates the special physical and chemical conditions for the solution and fractionation of the individual elements. In this the intensity of leaching and the kind of migration of the pervading solution are a function of the precipitation and evaporation conditions. As is well known, three main climatic zones are distinguished on the earth:

(*a*) The *humid* climatic zone (Europe belongs here) is characterized by the total amount of the annual precipitation being greater than that of the water being evaporated. This creates the most favourable conditions for the growth of vegetation. Weathering in these circumstances takes the form of chemical decay. Whereas the easily soluble substances are transported far away by running water, the substances of low solubility are carried downward within the soil together with the solutions resulting from the decomposition of the rocks. The removal of the bases results in pH-values lower than 7, the soil shows acid reaction (zone of leaching or oxidation). Some leached-substances are transported downward by the percolating solutions to be precipitated again in the increasingly neutral to alkaline zone (zone of enrichment or cementation; Fig. 43). This zone of enrichment or cementation may acquire an extraordinary importance for the mining of ores. The humid climatic zone is divided into two belts: the tropical-humid belt (precipitation up to 10,000 mm), and the temperate-humid belt (precipitation up to 4000 mm).

(b) The *arid* climatic zone is characterized by the total amount of the annual precipitates being very small (<250 mm). Precipitation is concentrated in a very short time during which rainfall is intense. Physical weathering becomes of great importance. The solutions of decomposed rocks, originated in the course of single violent rain-storms, soon begin to move upward by capillary force due to evaporation. At the earth's surface the solvents evaporate and the substances are redeposited in the form of efflorescent salts and concretions within the upper soil

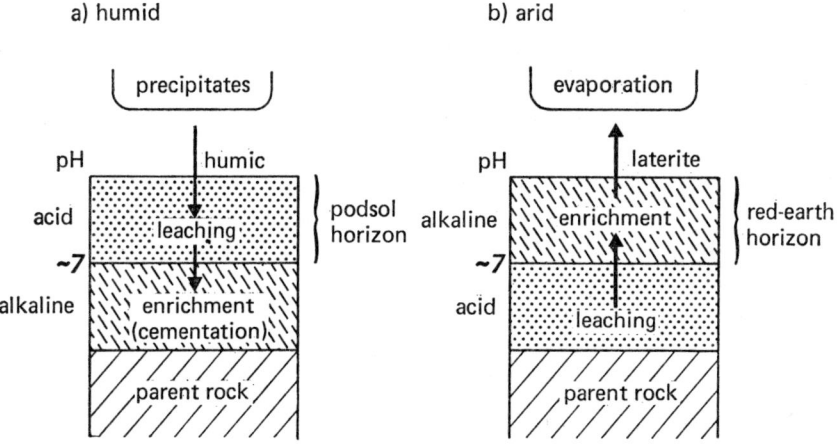

FIG. 43. Soil sections showing dependency on climate.

layers. These now possess an alkaline reaction. The outer crust instead of being leached is enriched by new precipitates (zone of enrichment; Fig. 43). The arid climatic zones form two belts, each of which separates, in both hemispheres, the tropical-humid belt from the temperate-humid belt. There are, of course, no sharp boundaries between these different climatic zones, and intermediate regions exist with semi-arid or semi-humid climates, respectively.

(c) The *nival* climatic zone comprises all areas, where precipitation is mostly in the form of snow and changes take place through glacier snow and granular ice to glacier ice. The zone of nival climate mainly includes both pole caps (approximately bounded by the Polar Circles) and high mountainous regions above the snow limit. It is remarkable that the total amount of annual precipitation is about the same as that of the arid climate (<250 mm). Chemical weathering is ineffective but physical disruption by frost-shattering can be important. The products of decomposition are transported by the glacier ice.

2.2.2. Transportation and deposition

Weathering prepares and begins the separation of the material constituents. The physically disintegrated or chemically dissolved rock material of the earth's surface may remain in the place of its formation and change into soil or it may be subjected to removal by erosion and transportation. The most important *agents of transportation* are water, wind, ice and gravity. The total effect of these exogenetic forces increases considerably, if they operate in combination. Which factors will be most effective in any region depends principally on climate and morphology.

Erosion or denudation is always followed by *deposition* of the transported material. The *causes of deposition* of the products of rock decomposition may be of both mechanical and physico-chemical nature.

Mechanical causes generally involve a reduction in the transporting force. This is brought about by a decrease of the surface-gradient and/or a lowering of the capacity of the transporting medium. Transportation and deposition bring about a natural sorting of the rock and mineral fragments.

Physico-chemical causes: In contrast to magmatic residual liquids *pressure and temperature* play scarcely any role in this context. The temperature is far below the boiling point of water and the pressure mostly amounts to only 1 atm. Locally, *turbulence* of the solutions may be of importance for the precipitation of the dissolved substances, since it may cause dissolved gases (for instance CO_2) to be given off. Also *evaporation* of the solvent may be very important for example, in lakes and isolated sea-basins under arid climatic conditions where it brings about the deposition of salts. Changes in the *pH-value* may cause the precipitation of dissolved substances. Their importance has been touched upon already when treating the solution of Al, Fe and Si (p. 65). Furthermore, changes of the *Eh-value* (redox potential) may play an important part in determining the kind of segregations formed. With colloidal solutions also *electrochemical* processes may bring about precipitation (mutual deflocculation of electrically positive or negative gels). If colloidal solutions or suspensions come into contact with a medium with a high *electrolyte content* precipitation also takes place. Thus many colloids are protected from deflocculation by so-called 'protective colloids' (for example humic substances). If these protective colloids are destroyed, coagulation and precipitation of the sols will take place at once. Finally, also *biochemical causes* may have great importance in causing precipitation. Among these bacteria may play a vital role in causing precipitation from solutions. The groups most important for the formation of ore deposits are lime-depositing bacteria, iron bacteria and sulphur (anaerobic) bacteria.

In the formation of sedimentary deposits there is no parallelism between the deposit-forming processes and, for example, decreasing pressure and temperature, as is the case in the formation of magmatic deposits. Therefore, a clear genetic classification is not easily set up. As an important criterion for classification the *distance of transportation* may be considered. The distance of transportation corresponds roughly to a subdivision based upon the *place of deposition*. We distinguish here between deposition or precipitation on the land (terrestrial deposits), in inland waters or fresh waters (limnic deposits), and in the sea (marine deposits). Operating as media of precipitation in the terrestrial regime we have ground and percolating water, in the limnic regime, river and lake water, and in the marine regime, sea water. Different kinds of deposits may originate in each of these media.

Transportation and deposition bring about preferential enrichment of certain elements. These elements may be united in groups of substances, much like the elements separating out together in practical chemical experiments. According to this principle a classification into the following groups results (Fig. 44):

I. Detrital sediments } mainly mechanical deposition ⟶ clastic sediments or clastic rocks

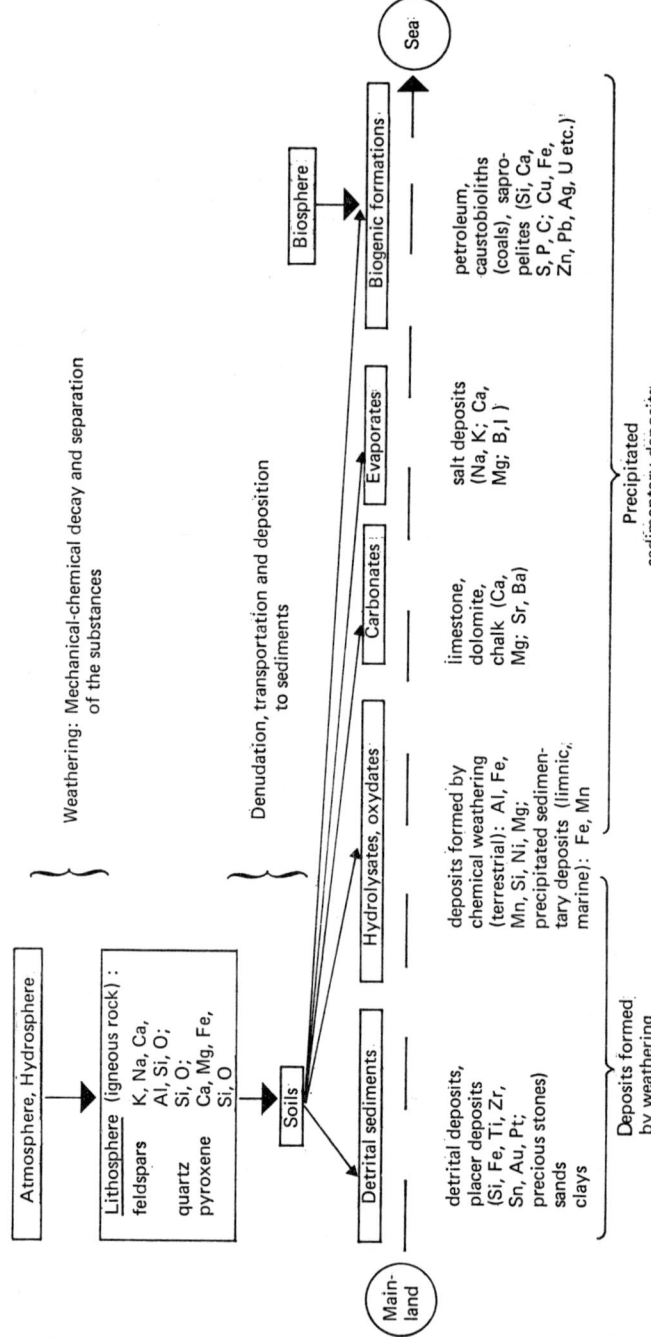

Fig. 44. The effects of transportation and deposition on the chemical composition and formation of sediments.

II. Hydrolysates, oxidates (partly residues in a wider sense)
III. Carbonates (limestones and dolomites in part)
IV. Evaporites

⎫ mainly chemical precipitation ⟶ chemical sediments

V. Biogenic rocks (limestones and siliceous rocks in part; phosphates, caustobioliths, sapropelites)

mainly biochemical precipitation ⟶ organic sediments

The various influences in the exogenetic cycle produce, of course, frequent mixing, so that the individual material groups are not always distinctly developed. This classification nevertheless provides a framework, into which all occurrences may be reasonably incorporated if the more detailed circumstances of formation are taken into account.

2.2.2.1. *Detrital sediments*

These are mainly products of physical weathering, that is clastic or fragmental sediments which are transported mechanically as solid constituents and are then deposited. As mentioned above, the separation of the clastic material according to grain size and density by gravity, water and wind represents a natural mineral dressing process. This causes, in the residual fraction, characteristic enrichment of certain mechanically and chemically stable minerals with the elements Si (quartz), Ti, Fe (rutile, ilmenite, magnetite), Zr (zircon), Sn (cassiterite), P (apatite), Au, Pt, precious stones (garnet, tourmaline) and others. The most important feature of these mechanical sediments are the grain-size and particle-size distributions, which are the result of transportation differing in distance and duration or of different transporting forces. The clastic sediments are divided into groups of different particle size (Fig. 45):

(a) *Psephites* (or *Rudites*) with particle diameters >2 mm (as loose material: gravels, pebbles, cobbles and boulders; as solidified clastic rock: breccia, conglomerate).

(b) *Psammites* (or *Arenites*) with grain diameters of 2-0·02 mm (loose rock: sands, placer deposits; consolidated: sandstone, arkose, quartzite, greywacke).

(c) *Pelites* (or *Lutite*) with a grain diameter <0·02 mm (loose rock: silt, muds, clay; consolidated: siltstone, shale, mudstone).

Useful concentrations of raw materials may occur in each of these

72 INTRODUCTION TO ORE DEPOSITS

groups. Among the psephites useful sedimentary deposits may occur due to the clastic material, derived from pre-existing deposits being further enriched by the sorting action of the transporting medium (detrital deposits).

The psephitic and especially the finer-grained psammitic materials

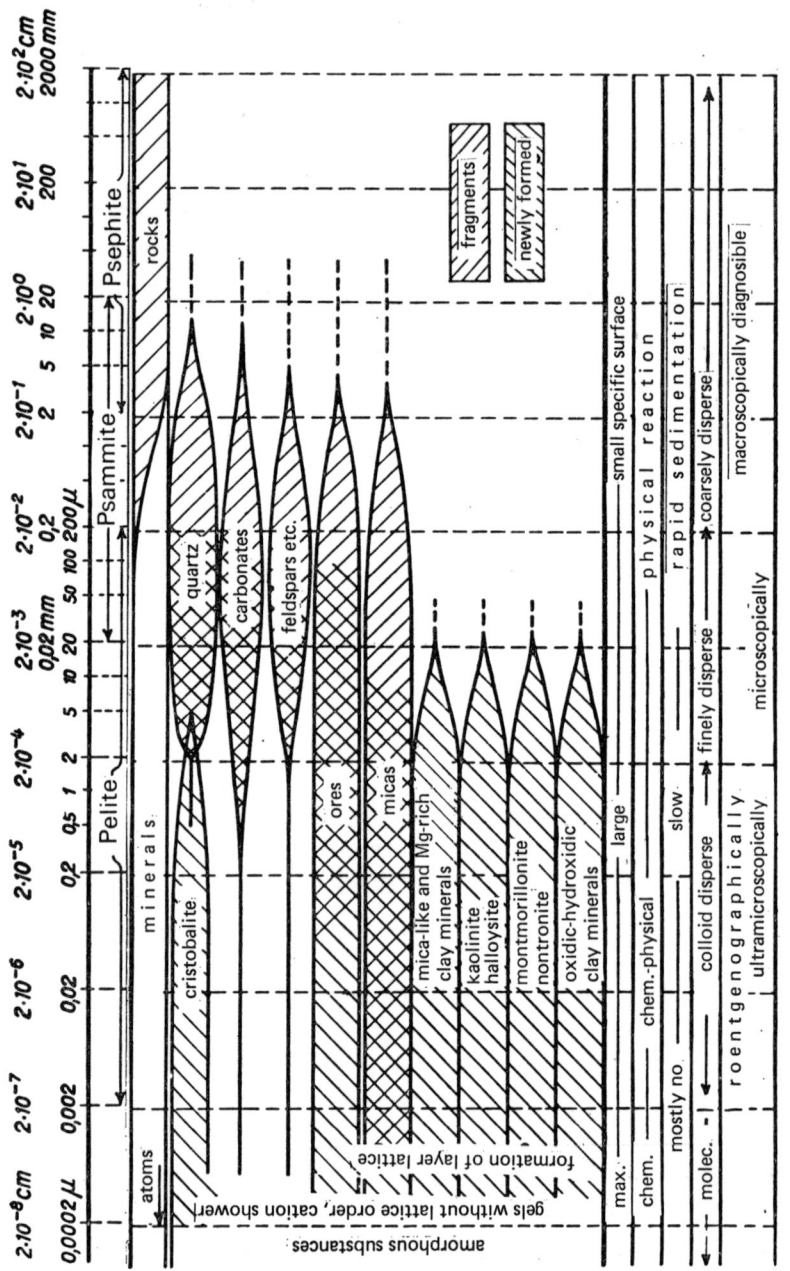

FIG. 45. Systematics of the detrital sediments (after Niggli, 1952).

are used in the *building industry* (as gravels, concrete, building sands, foundry sands, quartz sands; as sandstone for building purposes and sculpture, as quartzite for refractory material and so on). Sands, containing heavy minerals, are termed *placers* and may represent important deposits for the extraction of ores and precious stones. These include ilmenite sands, gold sands, cassiterite sands and garnet sands. The porosity of the sands or the sandstones and the frequent juxtaposition of psammites and sapropelites bring about the formation of oil sands and oil sandstones. Frequently alternating layers of coarser sands and finer bituminous clay may result in especially rich *petroleum deposits*. In this case the psammites form the reservoir rock.

Pelites represent a mixed association, containing besides the finest mineral particles (partly colloid-disperse particles) a large portion of *newly formed minerals*. The individual grains cannot be distinguished macroscopically.

Clays consist principally of layered minerals such as sericite, kaolinite, montmorillonite, illite. The clay minerals may be joined by still coarser sandy material or carbonates. An increase of the sand content in the clay results in loam, that of carbonate results in marl. Because of their plastic properties clays, loams and marls are used in *ceramics* (refractory clays, high-grade clays), in *brick-making* (brick clay, brick loam) and in the *cement industry* (cement marl, calcareous marl). Diagenetic changes may lead to a thoroughly compacted shale, which may be of use for making *roof slates*.

2.2.2.2. *Oxidates and hydrolysates (=hydrogelites)*

These differ from the group treated above in that there are no longer present undecomposed clastic solid constituents, but exclusively products of chemical decay and dissolution. Here we find the products of decomposition of pre-existing deposits, especially within the *zone of oxidation*. Within this group those elements are enriched which normally remain on the weathered land-surface, due to their generally lesser mobility. They are mainly the elements Al, Fe, Mn, Si, Ni, and Mg. These become precipitated from the finely dispersed or colloidal-dissolved state as hydroxides (of Al, Fe; e.g. laterite, bauxite), hydrated oxides (of Fe, Mn) or as hydrosilicates (of Si, Al, Ni, Mg; e.g. kaolinite).

Besides these near-source precipitates, which are chiefly confined to terrestrial deposits, some elements may be transported much farther under special circumstances. Fe and Mn especially may be transported very far in the form of carbonate ionic solution, or also hydroxide gel providing they are protected by organic colloids from deflocculation. They may then form deposits in limnic and marine areas. Si, too, may reach the sea in colloidal solution under certain conditions. There it is either deflocculated as silica gel (to form chert) or it serves in the building-up of several organisms (siliceous algae, radiolarians, sponges). The decay of the organisms leads to the formation of siliceous deposits (diatomites, radiolarites, spongiolites).

2.2.2.3. Carbonates

This group is enriched in the alkaline-earth elements Ca and Mg. Geologically, Ca- and Mg-carbonate deposition plays a considerable role. They may be of inorganic and organic origin. Their proportion is limestone:dolomite $= 2\cdot3:1$. Sr- and Ba-carbonates (strontianite and witherite) also occur as well as the sulphates, anhydrite, gypsum, celestite and barite. Limestones are often of economic importance as excellent building stones. Furthermore they find extensive use in the cement and metallurgical industries as well as in agriculture as a fertilizer (agricultural lime).

2.2.2.4. Evaporites

This group comprises all those mineral associations which form mainly by evaporation from marine solutions although compared with the carbonates they remain longer in solution. The formation of commercial salt deposits takes place with the precipitation especially of the alkali metals (Na, K). The most important preconditions for the formation of salt deposits are a sea basin, that has partly lost its connection with the ocean, and an arid climate, allowing a sufficient intensity of evaporation. A more recent suggestion, based on investigations along the shores of the Persian Gulf, is that precipitation may occur in intertidal areas and that isolation from the open ocean is not necessary (Shearman, 1966).

As compared with the saline deposits, the less frequently occuring terrestrial formations of salt reveal a higher proportion of Na-carbonate (alkaline lakes producing for example soda, $Na_2CO_3 \cdot 10H_2O$). Moreover, typical of this zone are sulphates ($Na_2SO_4 \cdot 10H_2O$, glauber salt), Ca-borate compounds ($NaCaB_5O_9 \cdot 8H_2O$, ulexite) and alkali-nitrate salts ($NaNO_3$, sodium nitrate; potash nitrate). The last require a very high redox potential, as exists for example in the high-lying Chilean pampas of the coastal cordillera with its saltpetre deposits.

2.2.2.5. Biogenic sediments

Within these groups the biosphere plays an important role as well as the atmosphere and hydrosphere. Thus, we have the participation of organisms in the formation of the siliceous rocks (siliceous algae, radiolarians and so on) and the carbonates (e.g. coral limestones, foraminifera, lime-depositing bacteria). In addition, iron and sulphur bacteria may be of importance in the formation of ore deposits. The anaerobic (sulphur) bacteria play an important role, together with the so-called desulphurizing bacteria and bacteria of decay, in the formation of sulphide ore deposits. Then, organic-rich shales can be petroleum parent rocks. Predominance of the sulphur bacteria, finally, may result in sulphate deposits (gypsum) or deposits of native sulphur.

Phosphorite is a very typical biochemical precipitate (normally this is $Ca_5F(PO_4)_3$ where OH^{2-} or CO_3^{2-} may substitute for F). Organisms at all stages of evolution take up P into their hard structure. Plants, too, store considerable amounts of dissolved phosphoric acid. In the process of organic decay soluble ammonium phosphate is produced, which is afterwards transformed, with lime or Al- or Fe-hydroxide, to form phosphorite or Al- or Fe-phosphate (wavellite or vivianite).

2.3. Formation of ore deposits of the exogenetic cycle

Based upon the above discussion the ore deposits of sedimentary origin may be divided into two great groups, namely 'ore deposits formed by weathering' and 'precipitated ore deposits'. The decisive factor for this division is the site of the events. Whereas ore deposits formed by weathering are mainly confined to the continents (terrestrial zone), the precipitated deposits occur in the terrestrial-limnic and the marine zones, thus involving longer distance of transportation of the dissolved matter.

2.3.1. Ore deposits formed by weathering

2.3.1.1. Ore deposits formed by mechanical weathering (residues)

Mechanical comminution and disintegration of the residue of weathering according to particle size and density represent a natural process of dressing. This process of sorting includes separation according to grain size and separation by specific gravity. The individual products of this process of sorting are again deposited at different places, partly on the land surface without any noteworthy transport (viz: 'detrital' deposits), partly in internal water reservoirs and in marine areas with longer distances of transportation (viz.: 'placer' deposits).

The *detrital deposits* consist mostly of coarse ore-detritus, mostly transported by sliding on slopes with more or less participation of running water. Certain ore concentrations may arise due to the removal of the specifically lighter and decomposable barren material. Naturally, such detrital deposits may be expected to form in zones of arid climate with great variations in temperature. If detritus reaches the surf zone of the sea further sorting of the sediments rich in ore will take place due to the action of the surf, especially during transgression of the sea (transgression conglomerates). Considerable ore concentrations may form as marine detrital deposits. An example is provided by the detrital iron-ore deposits near *Salzgitter* and *Peine-Ilsede* in the northwestern foreland of the Harz Mountains, which form the most important producers of iron in West Germany. These deposits consist of transgression conglomerates of the Cretaceous sea. The conglomerates consist of:

    I. rolled detritus of limonite;
    II. limonite- and siderite-oolites; and
    III. a ground mass (matrix) of clayey-sandy-limey nature.

The Fe is derived from clay-iron concretions of the underlying Jurassic layers.

Placer deposits. There are continuous transitions between the detrital deposits and the placer deposits. Some important placer minerals (with specific gravity) are:

Pt, Au	17·0
cassiterite	7·0
magnetite } monazite } ilmenite	5·0
diamond, garnet	3·5
basic silicates	3·0
(cf. feldspar, quartz	2·6)

Geologically, placers may be divided into 4 types:

(a) *Eluvial placer deposits.* These develop directly from detrital deposits and are therefore found in the immediate vicinity of primary deposits. They are formed from only slightly drifted residues, the lighter and decomposable barren material having been washed away and the heavier ore rubble remaining in place. Examples of eluvial placer deposits are, in part, the Pt-placers of the Urals (Nishni Tagil).

(b) *Fluviatile (alluvial) placers* represent the most important type of placer deposit. They develop from eluvial placers by brooks and rivers taking up and carrying off the residues of decomposition. Due to the continued movement of the water, separation takes place by specific gravity, the heavy minerals moving downward to the bottom of the river. Especially intensive enrichments are found where the flow velocity is decreased (pits in the river bed; at the inner side of sharp bends of the river and other change of cross-section; reductions of gradient at the places showing a confluence of tributaries; Fig. 46). It is noteworthy that for Pt and Au the process of concentration is not a purely mechanical one, but is associated with a kind of 'chemical purification'. This is recognizable by a decreasing content of Ag in Au and Ir, Os and the

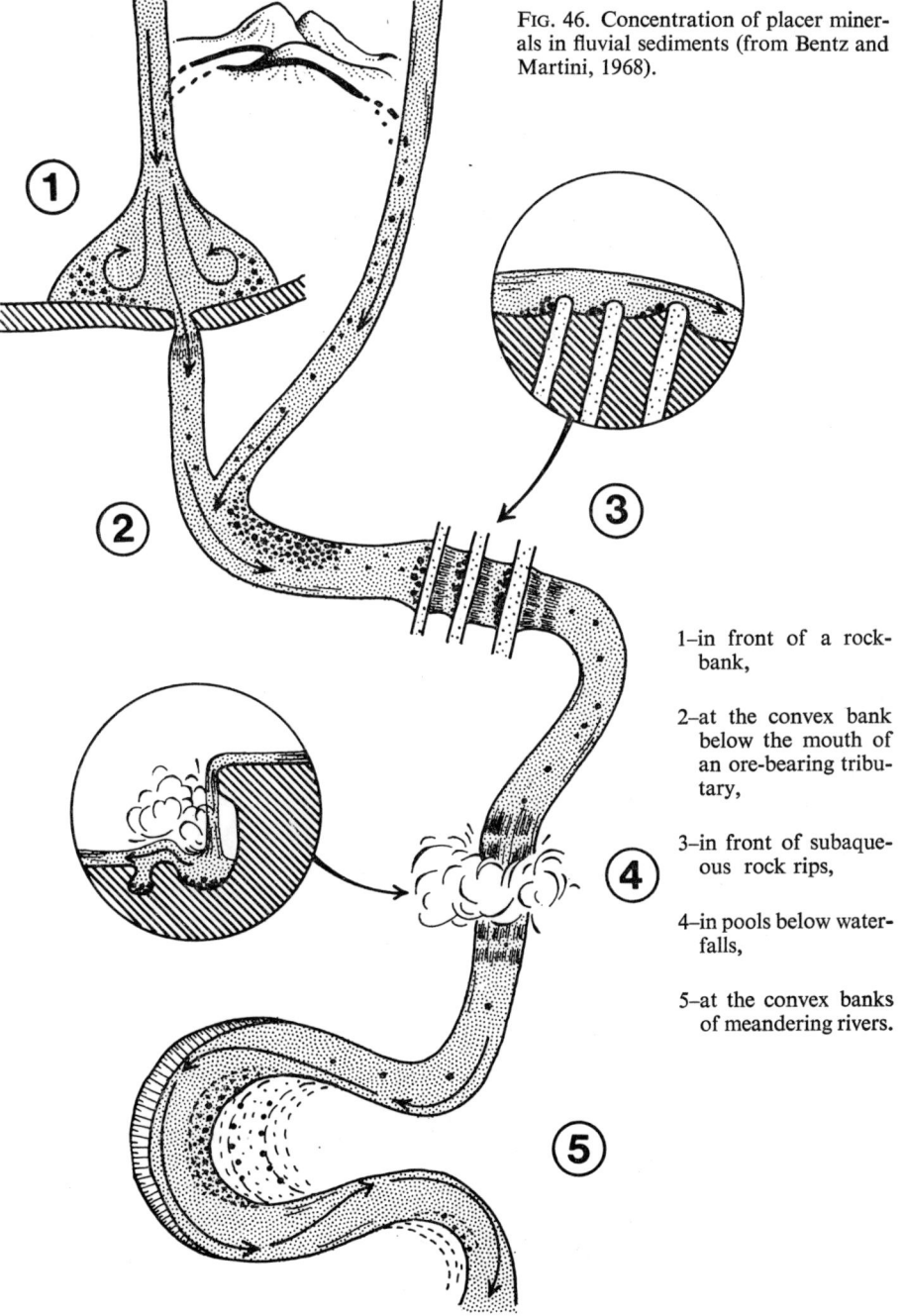

FIG. 46. Concentration of placer minerals in fluvial sediments (from Bentz and Martini, 1968).

1–in front of a rock-bank,

2–at the convex bank below the mouth of an ore-bearing tributary,

3–in front of subaqueous rock rips,

4–in pools below waterfalls,

5–at the convex banks of meandering rivers.

other transition metals in Pt (by leaching). Moreover, the fine gold spangles unite to form larger concretions and lumps ('nuggets'). Examples of commercially valuable fluviatile placers are the well-known *gold* placers of California, Alaska and Siberia (Upper Lena river) as well as the *tin* placers in the Erzgebirge and Indonesia (a most important tin producer).

(c) *Beach (marine) placers.* Placer minerals reaching the sea coast may be further enriched. The beach placers form preferentially near beach ridges that develop under the action of surf and tides. The flow at high tide and the surf waves throw the heavy minerals on to the shore, whereas the ebbing flow washes off the lighter constituents seaward. In this way very extensive and thick placer deposits often arise. Well-known examples of beach placers are the magnetite-ilmenite placers (western coast of Italy) as well as the remarkable monazite placers of Brazil and India. In addition, the rich diamond placers off the coast of Southwest Africa should be mentioned.

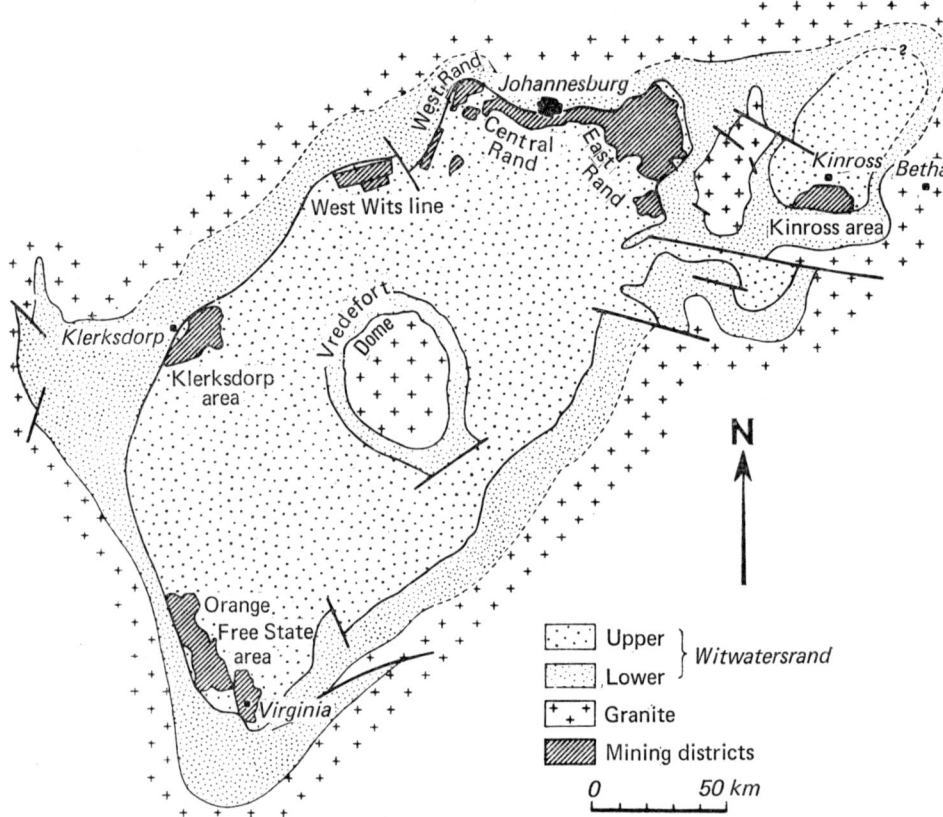

FIG. 47. Tectonic scheme of Witwatersrand basin with positions of mining areas (from Bentz and Martini, 1968).

(d) *Fossil placers* are, geologically, mostly very ancient placers, which have been lithified and partly sometimes thoroughly, metamorphosed. For example, the Precambrian gold placers of *Witwatersrand* in South Africa belong here (Fig. 47). The Witwatersrand is the world's largest gold district where more than one-third of the total production of gold is mined. The gold is found in several layers of conglomerates, embedded in a thick series of Algonkian quartzites. Owing to the presence of magnetite schist in the foot wall of the main ore horizon, geomagnetic measurements have been successfully made use of in exploration. The lateral extension of the gold layer is today proved to amount to about 250 km. Following the dip, mining has reached a depth of 3000 m. These are the deepest mines of the world and operations are possible because of the low geothermal gradient (1 °C per 130 m). The ore-carrying conglomerates consist of quartz rubble with particles of nut or egg-size and a quartzose cementing matrix with pyrite and gold metallization (at an average 10 ppm). This deposit is at the same time one of the largest uranium deposits of the world owing to accompanying uranium minerals. Similar deposits are also found in *Ghana* and *Brazil*.

2.3.1.2. *Ore deposits formed by chemical weathering*

These deposits are not formed by the enrichment of predominantly undecomposed primary minerals, but by reprecipitation of the products of chemical weathering.

(a) Occurrences due to weathering of *exposed deposits* (zones of oxidation and cementation, without considerable cation transport):

If ore deposits themselves are exposed to the decomposing influence of the agents of the atmosphere, voluminous metal displacements will take place. These displacements are most impressive in humid climates, where downward directed streams of percolating water are present due to the abundant rainfall. The zone of decomposition with ore deposits is divided into an upper zone of oxidation ('iron cap' or 'gossan'), which is characterized by intensive leaching processes, and a deeper-seated cementation zone rich in ore. The boundary between the two is approximately the ground-water level. The zone of cementation, finally, is followed by parts of the primary, unchanged deposit (Fig. 48).

In the *oxidation zone* the transforming agency is first of all the percolating water with oxygen, carbon dioxide, and salts dissolved in it. When the percolating water reaches the ground water, which lacks oxygen, its dissolving capacity becomes practically zero, because the ground water is saturated with the substances contained in its catchment area. Whereas in the uppermost part of the oxidation zone leaching

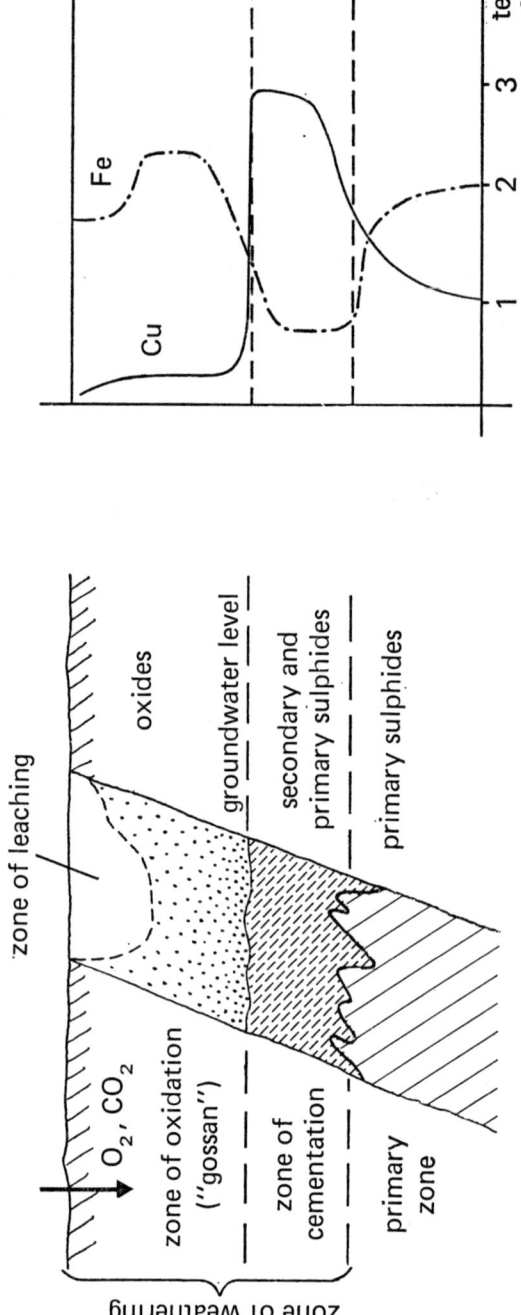

Fig. 48. Zones of oxidation and cementation, and occurrence of ores.

predominates, the metal oxides concentrate in the lower part. The depth of the oxidation zone depends on the ground-water level and may therefore vary between a few metres and some hundred metres.

In the *zone of cementation* the metal content of the sulphate solutions, derived from the oxidation zone, augments the unaffected primary sulphide ores. In compliance with the laws of the electrochemical series of the elements, the nobler metals, that is those having the higher potential, are precipitated on sulphides of less noble metals in the form of rich sulphides. The less noble metal, however, is then dissolved as sulphate (Electrochemical series: Au-Ag-Cu-Sn-Pb-Ni-Co-Fe-Zn-Mg-K). Since, for example, Fe lies far back in the electrochemical series, the common Fe sulphides react to precipitate many nobler metals:

$$4 \text{ FeS}_2 + 4 \text{ CuSO}_4 = 2 \text{ Cu}_2\text{S} + 4 \text{ FeSO}_4 + 3 \text{ S}_2$$

Commercially valuable natural concentrations of precious metals are thus produced (Fig. 48). The cementation zone unites in itself the metal content of the overlying oxidised and leached areas and that of parts of the primary deposit. Therefore it is of great importance for the mining industry. Almost all outcropping deposits of non-ferrous and precious metals provide examples of these processes; especially distinct oxidation-cementation zones are found in the copper deposits of Tsumeb (S.W. Africa) and Katanga (Congo), in the Pb-Zn-Ag deposits of Broken Hill (Zambia), and Cœur d'Alene (USA).

(*b*) *Laterite and bauxite deposits*. Aluminium deposits originate mainly by the decomposition of feldspars, which constitute more than 57% of the minerals of the earth's crust. In a *temperate-humid* climate Al_2O_3 and SiO_2 separate out simultaneously from the slightly acid solutions involved in rock decomposition to form together aluminium silicates (such as kaolinite). Thus, clay and china clay deposits are formed. In *tropical-humid* to *semi-arid* climates, however, preferential enrichment of Al and Fe takes place during the dry periods with neutral to alkaline reaction of the ground waters (SiO_2 is largely dissolved; Fig. 42). These enrichments are preferentially bound to the warm moist belt of the earth, characterized by an alternating climate with rainy and dry periods. The aluminium is first dissolved by acid, mostly carbon-dioxide bearing solutions, such as occur in the tropical rainy season. In the subsequent dry period the Al rises upward together with the ground water by capillary force and is precipitated at the surface by the alkaline ground waters as Al-hydroxide ($Al(OH)_3$, hydrargillite). On the other hand, the other constituents (Si, alkaline earth's, alkalies) remain in solution and are carried off. The encrustations and nodules of Al, having a yellow-red to intensively red colour in mixture with the coprecipitated Fe-hydrate, are commonly termed *laterites*. Depending on the composition of the decomposed rocks these may be mined, with very high contents of Al, as bauxite deposits or, with high contents of iron, as so-called 'lateritic iron ores'.

The workable *bauxite deposits* are mostly very extensive in area. Due to their mostly unconsolidated nature the lateritic cover of decomposed rocks is frequently eroded and redeposited in adjacent depressions. Accordingly, the bauxites are often divided into autochthonous bauxites or silicate bauxites and allochthonous bauxites or transported bauxites. Important examples of silicate-bauxite deposits are those of the Deccan Plateau/India (Fig. 49).

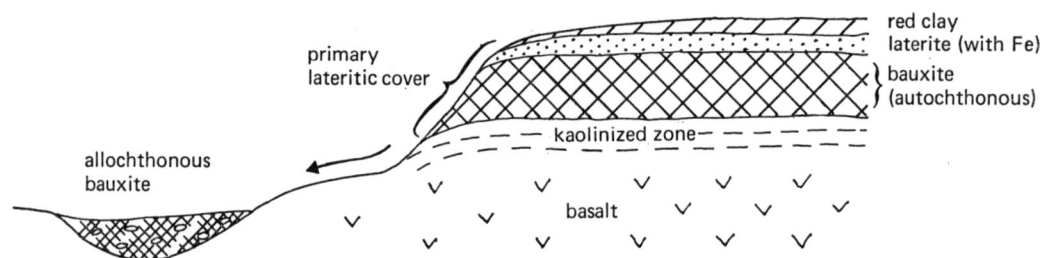

FIG. 49. Occurrence of silicate-bauxite deposits (Deccan type).

Beside the sheet-like bodies on silicate rocks bauxite deposits (lime bauxite) also occur in cavities of limestone and dolomites converted into karst topography. This is probably a residual material rich in aluminium, which was washed into the karst relief and became laterized immediately afterwards. The external shape of such deposits is adapted to the shape of the karst holes and therefore very variable (Fig. 50). The most important lime-bauxite deposits are chiefly confined to the folded carbonate rocks (Triassic, Cretaceous, Tertiary) of the countries round the Mediterranean Sea (e.g. Les Beaux, Southern France; Halimba and Gant, Hungary; Yugoslavia; Italy).

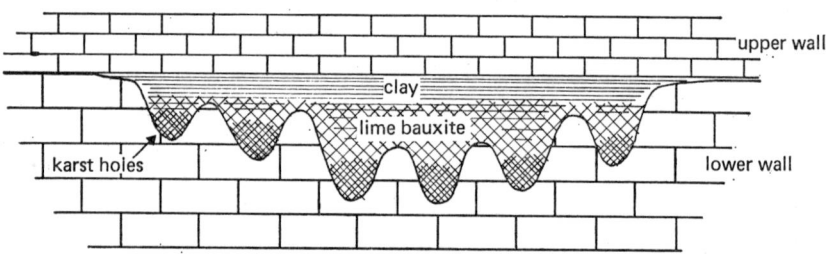

FIG. 50. Scheme showing the bedding of lime-bauxite deposits.

(c) The *laterite iron-ore* occurrences are mainly found as products of the atmospheric weathering of rocks high in iron (mostly basic igneous rocks: basalt, serpentinite, gabbro and others). The process of weathering corresponds to the formation of the bauxite. Also the geologic structure of these iron-ore deposits is similar to that of the bauxite deposits. The most important of these deposits lie in tropical areas in all continents (for example Conakry, Guinea; Mayari, Cuba).

(d) Under conditions of a temperate-humid climate the Fe goes into solution in slightly acid liquids resulting from rock decomposition (as Fe hydrosol). Also Mn is present in these conditions as hydroxide ($Mn(OH)_4$ in the form of a sol) and can be carried off too. Since the Fe hydrosol possesses a positive electric charge and the Mn hydrosol a negative one, the two can flocculate each other electrolytically to form mixed gels. This applies especially to limestone areas, where they can accumulate, like the lime bauxites, in trough-like karst holes. In this case the ores form irregular masses, mostly near the bottom of the karst cavities and are mostly separated from the surrounding rock by a narrow bed of clay (therefore water-retaining). Such ore types occur mainly in the karst districts of Middle Europe (e.g. W. Germany: Hunsrueck Mountains, Wuerttemberg), where they were intensively mined, especially in the Middle Ages.

Among these deposits may also be mentioned the so-called '*bean ore*' occurrences. These pisolitic limonites lie in a residual clay ('terra rossa') which is high in iron, in the form of rounded shell-like concretions. The formation of concretions was influenced here first of all by the absorbing action of the clay minerals. These ores too are mainly found in limestone caverns of old karst areas (for instance Swiss Jura, Swabian Alp). Characteristic of these ores is their abundant content of trace metals (V, Cu, Pb, Zn, Ni), which locally may form separate concretions.

(e) *Nickel silicate deposits.* In many ultrabasic rocks (peridotite, serpentinic and others) a small content of nickel is bound isomorphously to the main rock-forming mineral, olivine. This nickel content may become enriched with Fe in the process of lateritic weathering. In this case the largest Ni concentrations are mostly found on a level immediately below the lateritic iron-ore cover. This results in a stockwork-like structure: At the surface the lateritic iron ores form 'Red Rocks', below them the zone high in nickel occurs in 'Green Rocks' (Ni-Mg silicates, garnierite; often together with pure Mg silicates, talc, and SiO_2, chalcedony; in addition some Co and Cr is also concentrated). These nickel-bearing zones of rock decomposition (with up to 6% Ni) may go down to a depth of 30 m and are mined in open cuts. Farther

84 INTRODUCTION TO ORE DEPOSITS

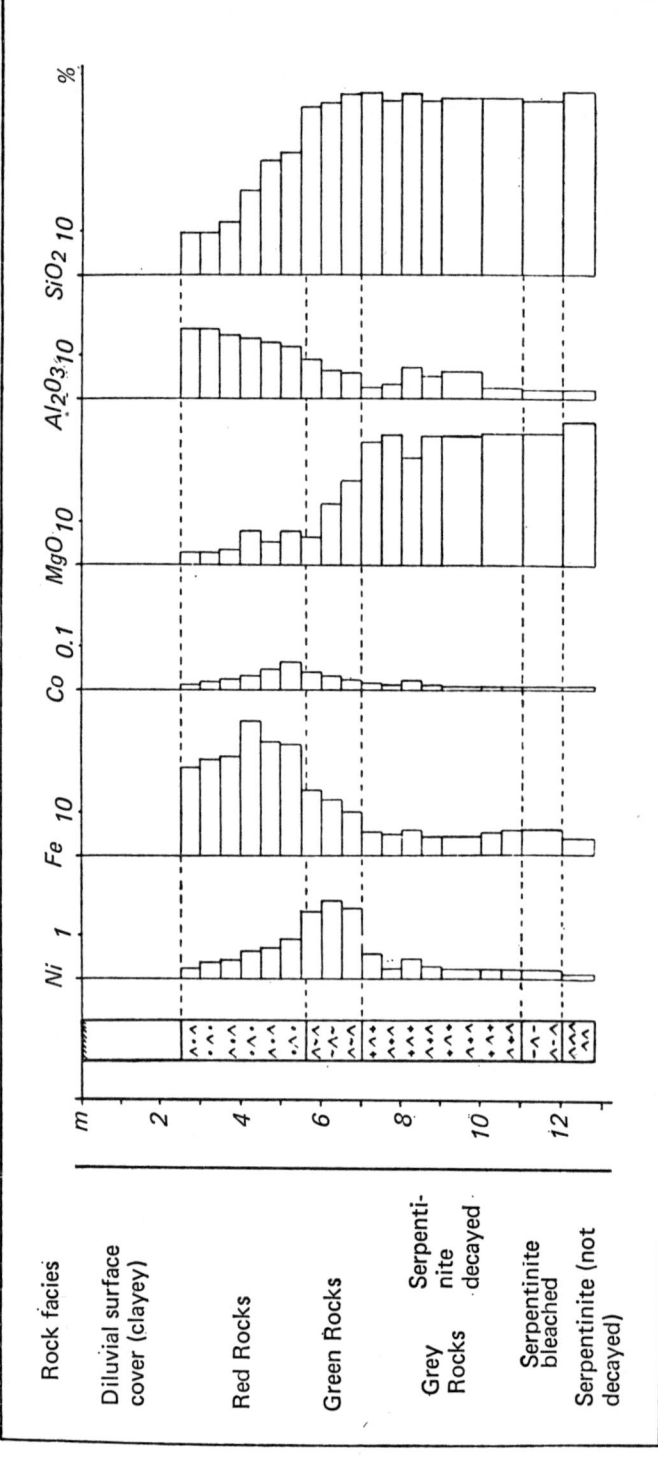

FIG 51. Schematic borehole cross-section from one of the deposits of the St. Egidien district (GDR).

down, this zone of decay then changes into bleached serpentine and, finally, undecomposed serpentine in the form of 'Grey Rocks'. An important deposit of this type is that near St. Egidien at the southern edge of the Saxon Granulite Mountains, GDR (Fig. 51).

In addition to the occurrence of St. Egidien such deposits are also found in *Czechoslovakia* (Křemže), in *Poland* (Lower Silesia) and in the *Soviet Union* (Orsk and Chalilovo, Southern Urals). Very important deposits of nickel silicate are also those of *New Caledonia* which, after Sudbury, are placed second in the world production of nickel. One-third of the island is occupied by Tertiary, serpentinized peridotites, from which the nickel ores (mainly garnierite) are mined in hundreds of open cuts (Figs. 52 and 53). The average contents are about 3% nickel and 2% cobalt. The cobalt, always accompanying the nickel, is present in the form of asbolan (Co-Mn oxide). Further large deposits are situated in *Cuba* and the *Philippines*.

(*f*) *Magnesites and phosphates formed by weathering.* It should also be mentioned that in the process of lateritic weathering pure precipitates of magnesite (gelatinous magnesite) and on apatite-carrying rocks the formation of phosphates are possible.

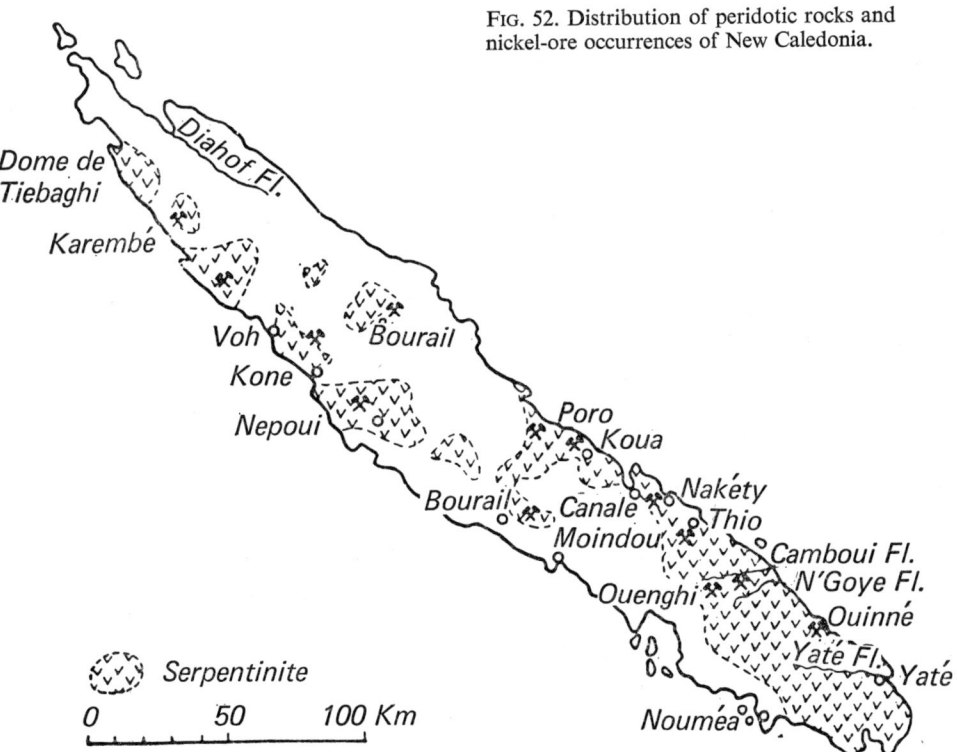

Fig. 52. Distribution of peridotic rocks and nickel-ore occurrences of New Caledonia.

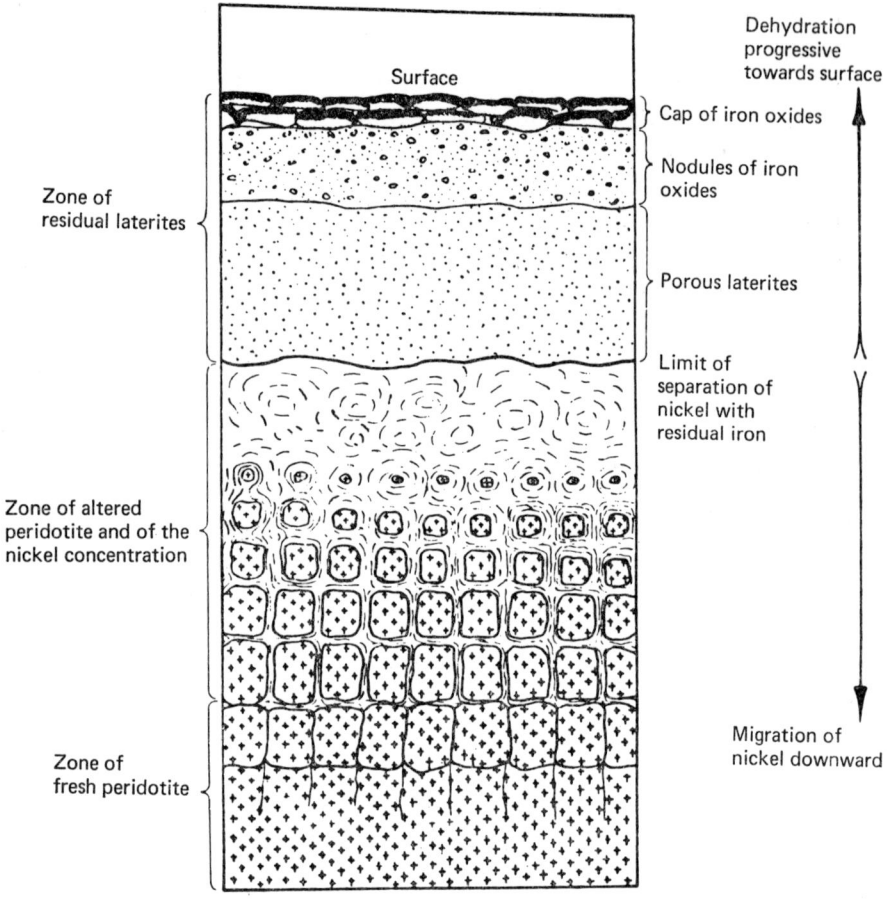

Fig. 53. Typical section through nickeliferous laterite deposits, New Caledonia (from Park and MacDiarmid, 1964).

2.3.2. Precipitated sedimentary ore deposits

Under certain circumstances some metallic ions are left in solution after weathering and remain so over variable distances of transportation until precipitated in arid basins, in inland waters (with humid climate) or in the sea.

2.3.2.1. *Precipitated deposits with medium-range transportation of cations* (*in arid basins and in inland lakes, rivers and ground water*)

(a) *Arid deposits formed by concentration (red-bed type).* In arid climates, due to the prevailing physical-chemical weathering, large accumulations of detritals are formed. These are spread over inland basins during short but violent rainy periods.

In these intercontinental depressions, mostly without outlet, the detrital material normally shows some lateral and vertical size-sorting. Whereas the margins and base of the arid detrital materials consists of relatively poorly sorted materials (fanglomerates), the upper and central parts consist of finer-grained and better sorted sediments (sandstones, siltstones and mudstones). These rocks are often distinguished by a red colour and are commonly referred to as 'red beds'. If primary ore deposits exist in the adjacent uplands or in the ground below the depressions, slightly concentrated metal solutions may be washed in (as sulphate or chloride solutions). This may happen directly or by groundwater movement and results in a syngenetic or epigenetic metallization forming so-called infiltration deposits. The metals are precipitated from the vadose water, occurring between the water-table and ground surface waters, which evaporate readily under arid climatic conditions to form sulphides, carbonates or oxides.

Ore precipitation can take place by *cation exchange* between the iron contained in the clay minerals and the nobler metals. In addition the *absorbing action* of the clay minerals themselves may have had a certain influence. Further, *plant material* included in the detritus is assumed to contribute to the formation of the sulphides by its strong reducing action on the solutions. Also extensive action of *bacteria* (sulphur bacteria) may favour ore precipitation. Therefore this type of deposit often reveals transitions to the deposits of the so-called '*sulphur cycle*' (see chapter 2.3.2.3.). Due to the short but violent storms reworking of the ore occurrences often takes place and this is frequently associated with considerable enrichment. The ores themselves are confined to bleached zones inside the red beds. They occur mostly as layered impregnations (up to 1 m thick) at certain levels, but are continuous only over short distances. The ore may occur irregularly in different levels of the thick detrital accumulations. Owing to the irregularity of occurrence, development and extraction of such ore deposits by mining are usually very complicated.

The important elements enriched in these arid detrital accumulations besides iron, are copper, lead, silver, vanadium and uranium. In places also alkalies and alkaline earths may be concentrated, that is as terrestrial salt deposits.

Important commercial ore concentrations of this category are first of all deposits of copper. These are found in many places on the earth. Such copper concentrations occur in the arid Permian sediments of the *Lower-Silesian-Bohemian basin*. Smaller Cu-impregnations are also found in the Bunter Sandstone on the border between the *Saar Territory and Lorraine* and have been partly mined. Also the *Permian* sandstone of the western foreland of the *Urals* (near the town of Perm) includes extensive impregnation horizons containing in places 2-6% copper. The copper content decreases gradually with increasing distance from the Urals which was the source-area. Important Cu-impregnation deposits in the Soviet Union occur in the Carboniferous sediments of *Kazakhstan* (for instance, Dzheskasgan district). A 900 m-thick sandstone series contains several large lenticular zones of impregnation (7 main ore horizons, preferentially confined to gray calcareous sandstones, rather than inter-bedded red, clayey sandstones; Fig. 54). Metallization (Cu sulphides, small amounts of PbS, baryte) in part also occurs in zones of tectonic disturbance. The copper content runs at about 1·5% and the metallized area comprises about 100 km². Reserves are estimated at about 3·5 million tons of metal. Classic red-bed deposits are situated in the west of the USA (Utah, Colorado, Arizona, New Mexico). Here a thick series of red sandstone (Upper Carboniferous to Jurassic) was developed. In different stages of it there appear Cu-impregnation deposits, whose metal content was probably derived from the surrounding basement eminences.

Vanadium also is concentrated in arid deposits of the red-bed type (for example in many deposits of Rotliegendes and Bunter age). Vanadium occurs mainly as carnotite (vanadate of uranium) in the Jurassic sandstones of the *Colorado Plateau* and east Utah, USA. Typical of these sandstones are thin seams of coal and carbonized drift wood and the carbonaceous material, acting as a reducing agent, causes enrichment of the ores. V-U metallization occurs in several horizons of impregnation. Probably the two elements were precipitated from the groundwater of the arid basins. The ore-bearing sandstone contains up to 5% V_2O_5 and 3% UO_3. These deposits are today's greatest uranium producer of the USA (Fig. 55).

(*b*) *Limonitic soils, bog iron-ores and lake ores.* Under reducing conditions, iron is readily soluble in CO_2-containing water as iron bicarbonate. Thus, for example, in groundwater (free of O_2) considerable amounts of iron bicarbonates or iron humates can be transported. If such solutions reach the surface together with the groundwater, contact with oxygen takes place followed by precipitation of iron carbonate or iron hydroxide respectively, CO_2 being liberated. In this, two minor types of iron-ore deposits may form, first limonite meadow ores or bog-iron ores and second lake-iron ores (Fig. 56). Precipitation of iron hydroxide may also take place due to the action of bacteria (iron bacteria). Mn may also be precipitated under similar

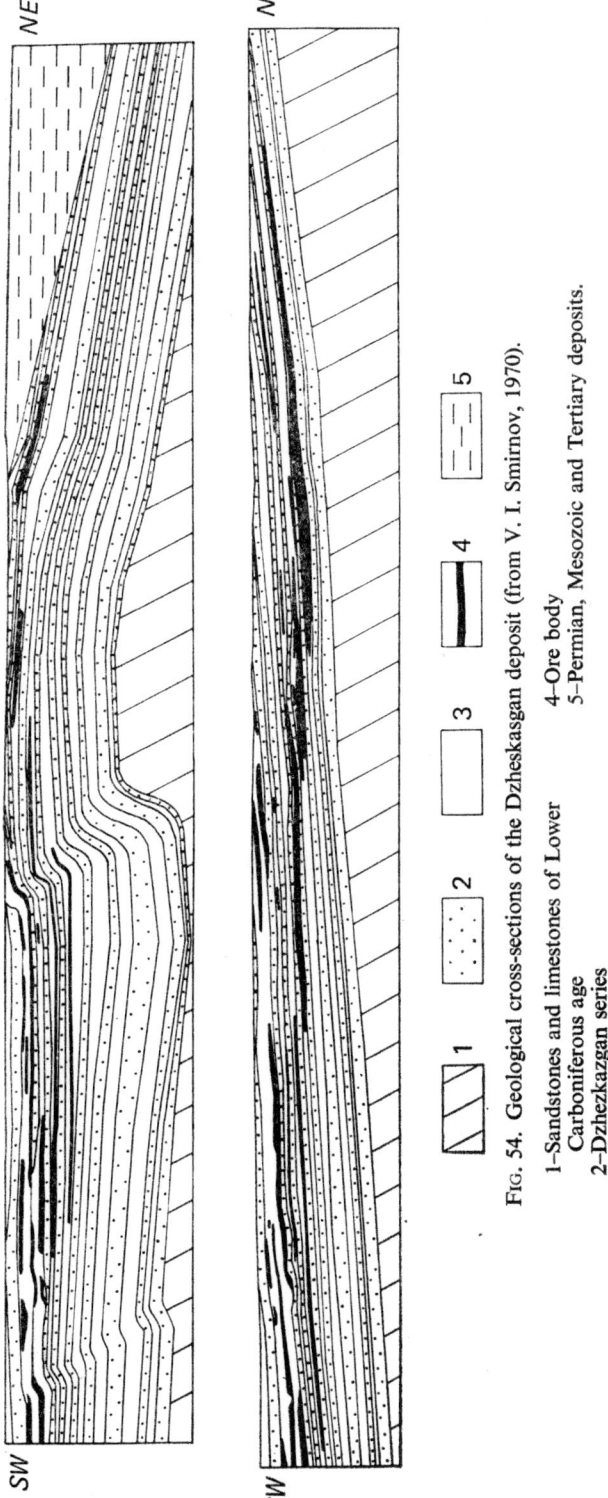

FIG. 54. Geological cross-sections of the Dzheskasgan deposit (from V. I. Smirnov, 1970).

1–Sandstones and limestones of Lower Carboniferous age
2–Dzhezkazgan series
3–Ore-bearing grey arkoses
4–Ore body
5–Permian, Mesozoic and Tertiary deposits.

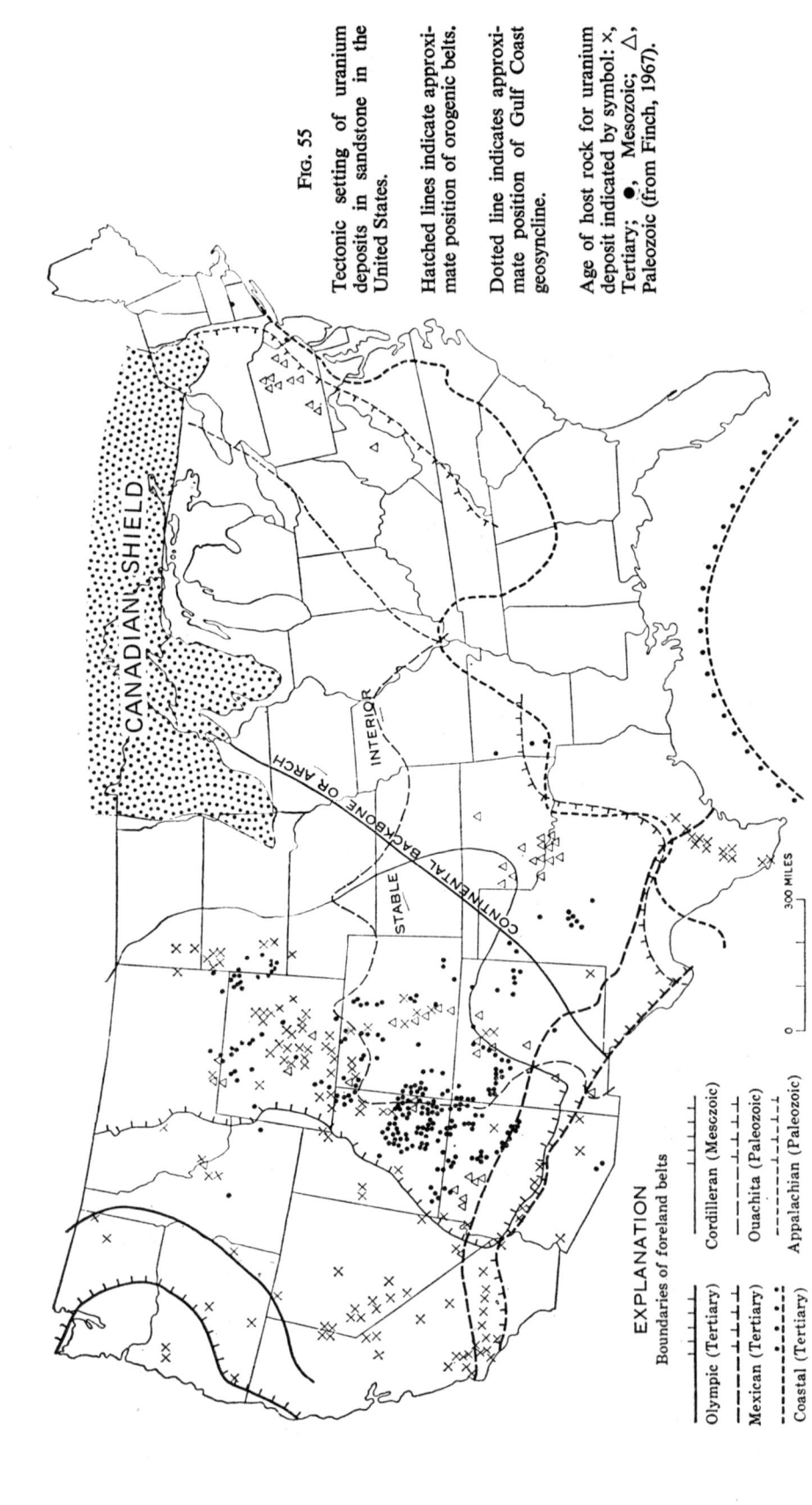

FIG. 55

Tectonic setting of uranium deposits in sandstone in the United States.

Hatched lines indicate approximate position of orogenic belts.

Dotted line indicates approximate position of Gulf Coast geosyncline.

Age of host rock for uranium deposit indicated by symbol: ×, Tertiary; ●, Mesozoic; △, Paleozoic (from Finch, 1967).

conditions. Remarkable occurrences of limonite meadow ore and bog-iron ore are found in *Jutland*, Denmark, as well as in the *Campine district*, Belgium. Deposits of lake-iron ores are also found in Finland and Sweden.

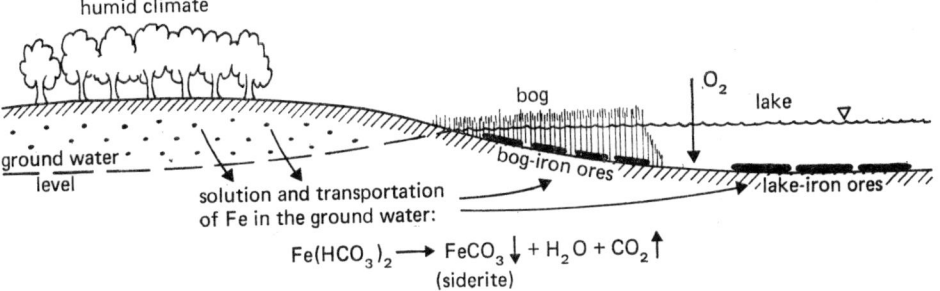

FIG. 56. Formation of the bog- and lake-iron ores.

2.3.2.2. *Precipitated sedimentary deposits with distant transportation of the cations* (*in epicontinental sea sediments*)

Since the sea normally is one of those terrestrial substances lowest in metal content (for example $2 \times 10^{-7}\%$ Fe) it seems astonishing that nevertheless enormous valuable concentrations of metals, especially iron, have formed under marine conditions. A special series of circumstances is necessary to bring about the formation of workable ore concentrations:

(*a*) *Origin*. There must have been present in the coastal area and on the continent sufficient rocks high in Fe (say older iron deposits) to ensure on weathering, a continuous supply of iron-bearing solutions to the marine area of sedimentation. Besides the origin of the constituents from solutions of rock decomposition from the mainland, H. Borchert (1964) also suggests submarine weathering as a source for part of these sedimentary iron deposits. The removal of the Fe from the sea bottom is assumed to be brought about by deep CO_2-bearing sea water. Furthermore, submarine hydrothermal supply of the metallic ions in the process of geosynclinal volcanism is possible.

(*b*) *Weathering*. On the mainland, vegetation of extensive bogs and swamp forests must have been present during times of intensive weathering (tropical-humid climate). The *humic waters* or organic-rich water first removes the iron extensively from the rocks to transport it then as Fe organic complex, Fe-bicarbonate, or also as Fe-hydroxide plus a protective colloid into the sea (the formation of this type of deposit type was therefore hardly possible prior to the Carboniferous because of the insignificant land vegetation).

(c) *Transportation.* Transportation over long distances was possible by the fact that the $Fe(OH)_3$-sol particles could be prevented from premature flocculation by the addition of certain protective colloids. As protective colloids for the slightly positively charged Fe hydrosol, derivates of humic acid, which are strongly negatively charged might be cited. Thus a great part of the Fe could be transported far into the sea as a protected sol. The Fe may also be transported in smaller amounts in the form of humate or as Fe-hydroxide bound to clay minerals by adsorption. In the end the protective colloids become decomposed, by the electrolytes (Na^+, K^+) present in the sea water and the action of bacteria, and the iron coagulates.

(d) *Precipitation and sedimentation.* The iron separates out in the sedimentary deposits as $Fe(OH)_3$, $FeCO_3$ or iron silicate. Which of these minerals forms depends largely on the different redox potentials (Eh-value) in the different parts of the ocean (the pH-value in the sea is subjected to only slight variations). In the solution and reprecipitation of the iron, Borchert (1964) distinguishes several areas of formation (Fig. 57): In the immediate coastal area O_2 is abundant in the sea water; therefore the redox potential is positive. This we might call the oxygen zone. The product separated out first of all in the oxygen zone, is *Fe hydroxide* (besides Al hydroxide and colloidal SiO_2). Consequently, limonite oolites form predominantly here partly intermixed with mainland detritus. With increasing distance from the coast the develop-

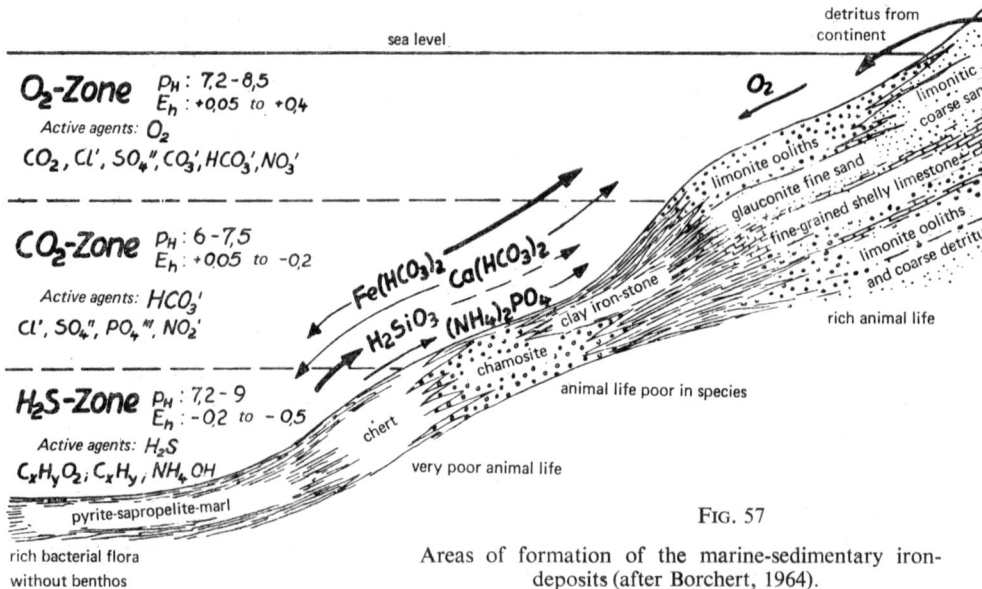

FIG. 57

Areas of formation of the marine-sedimentary iron-deposits (after Borchert, 1964).

ment of CO_2 increases due to intensified decay of organic matter thus giving rise to a slightly reducing environment (the carbonic-acid zone). Fe^{3+} is reduced to Fe^{2+} and *ferrous carbonate* is separated out (in part as oolites). The siderite occurs mostly in the form of finely crystalline concretions embedded in the clayey sediments (clay ironstone). With further decrease in Eh-value the Fe^{2+} ions then combine with the colloidally dissolved SiO_2 to form *iron-silicate gels* (chamosite, partly as gel globules, partly as true oolites). In the succeeding zone of stagnant water, strongly reducing conditions prevail, due to abundant decayed organic matter (Eh-value negative). Here H_2S is generated and the iron is precipitated as *iron sulphide*.

(*e*) *Formation of oolite*. For the formation of a deposit of useful minerals to take place the supply of clastic sediments (sand, clay) and the marine precipitation of carbonates (organic and inorganic) must be less than the precipitation of iron. The ores occur mostly as oolitic ores. These form by the colloidal solutions coagulating on small suspended foreign bodies (quartz grains, fossil fragments, bacteria and so on). Concentric shells form of changing composition, positively charged Fe- and Al-hydrogels and negatively charged SiO_2-gel, coagulating each other in layers. The deposition of these round shell-like oolites takes place when they have become so heavy that their settling velocity no longer allows them to be maintained in suspension with the existing movement and density of the water. This process of settling leads to the oolites of the individual ore horizons being distinguished by a relatively uniform size (0·3-2 mm). They are then mostly embedded in calcareous-clayey material which always contains some iron as well (Fig. 58). The oolites, deposited on the sea bottom, may suffer further transportation; they may be washed together and agzin sorted. They can also be partly crushed and a new crust may form over the fragments.

2.3.2.2.1. *Oolitic iron-ore deposits* (*minette type*)

Where the above-mentioned requirements were met, the formation of oolitic iron-ore deposits could take place. Palaeographically, these deposits were probably associated with largely isolated parts of the sea, where sufficiently high iron concentrations could arise. As compared with the oolitic, chamosite-thuringite deposits, the limonitic-oolitic deposits contain much less Ti, V, Cr, Cu, Pb and Zn and are in general very poor in trace elements. Since, however, in their formative environment organisms were very abundant the oolitic limonite ores possess relatively high P-contents (0·6-1·0%). Because of their lime content these ores can be smelted although the high P content necessitates that a special method known as the 'basic Thomas process' be used.

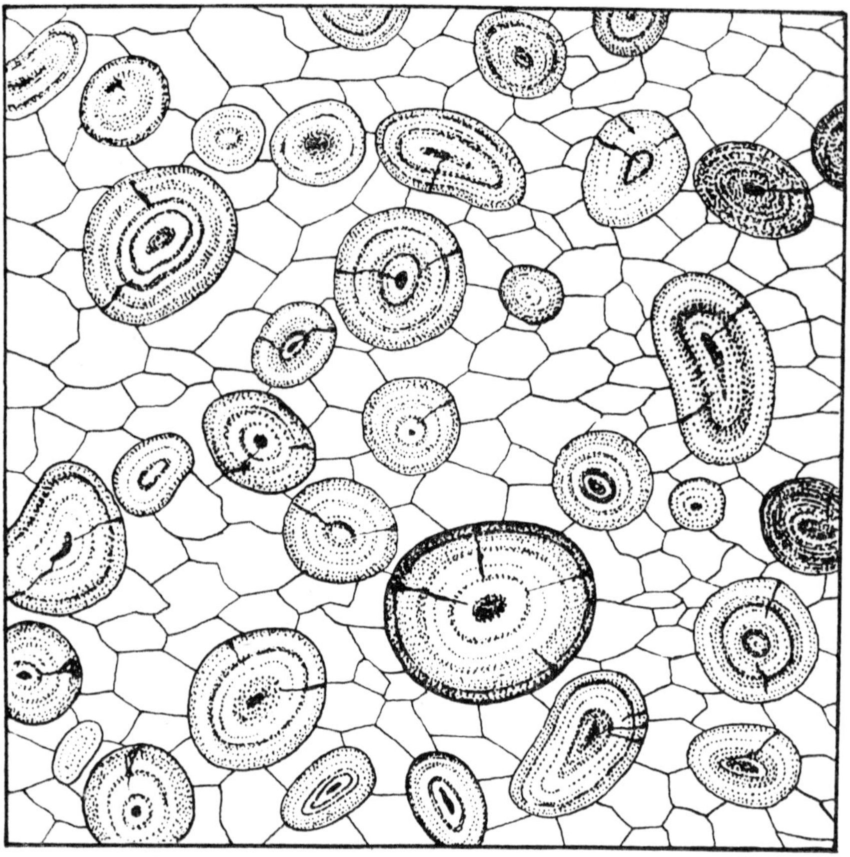

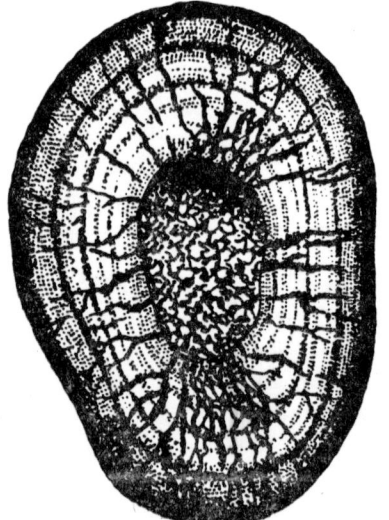

(a) Oolite grains in a granular matrix.

FIG. 58

(b) Structure of an iron-oolite grain with shell-like texture and cemented cracks (from Niggli, 1952).

This type of ore deposit is mainly met with in the Jurassic rocks of Western and Central Europe (Lorraine-Luxemborg, Baden-Württemberg).

Lorraine-Luxemborg. Here the limonite ores were mainly developed in Lower Dogger (Middle Jurassic) rocks. There are remarkable lenticular ore deposits, 8 to 10 of them lying one above the other, their thicknesses varying between 3 and 9 m. The iron content varies from 30 to 40% and the ratio $CaO:SiO_2$ is 1·4:1 giving a 'self-fluxing ore' in the furnace. The contents of Mn are very small. The ores are mainly extracted by open-cast mining. The reserves are estimated at about 15,000 million tons.

Baden-Württemberg. These are similar ores, also in the Dogger sequence, but lower in Fe (25% Fe) and higher in SiO_2. Further occurrences are found in *Franken* (northeast of Nürnberg) and in *north Germany* (in Liassic rocks). Oolitic limonite ores also occur in central England (Liassic, Dogger), in the Ukraine and North Africa (Algeria and in Tunisia, Eocene; Egypt, Senonian). It may be finally stated that this type of deposit is always associated with geological epochs of large-scale transgressions of the sea and they tend to be widespread. It may be added that many oolitic iron-ore deposits consist only of *chamosite-thuringite* (ferrous silicates rich in water and with a greenish-grey to green-black colour) or *hematite*, respectively. Formation of oolitic chamosite iron ores took place, in the region of Thuringia, GDR, within rocks of Lower Silurian age. Further important chamosite deposits are located in the Lower Silurian series of the *Prague Trough,* Czechoslovakia (Nučice, Kyšice) and in *Normandy* and *Brittany* (where France has 1600 million tons of ore reserves). Because of the frequent close association of the chamosite ores with the rocks of the submarine diabase volcanism, a magmatic origin has been assumed recently for a large part of these deposits. There should be mentioned also the Upper-Silurian *Clinton ores* of Birmingham, Alabama (two seam-like iron-ore deposits consisting of oolitic and clastic hematite; contents have 35% Fe, reserves 1500 million tons). These occurrences are, after the Lake Superior district, the second-largest iron-ore district of the USA.

2.3.2.2.2. *Oolitic deposits of manganese ore (Tshiaturi type)*

Mn-ores may form by marine-sedimentary processes similar to those responsible for the iron-ore deposits. Nevertheless, these two metals occur more or less separated from each other. This seems to be caused by the higher Mn contents in the weathering source areas, on the one hand; on the other, the separation of Mn and Fe in rock decomposition and precipitation may be explained by the fact already mentioned that the Fe hydroxide may be deposited considerably earlier than the Mn hydroxide (Fe^{3+} at pH =3, acid environment, e.g. in bog water; Mn^{4+} only at pH =8). Thus a spatial separation of both elements may take place. Whereas, for example, the $Fe(OH)_3$ may be assumed to have been precipitated near coastal areas, the Mn hydroxides reached parts of the sea more distant from the coast. With very extensive epicontinental seas the formation of purely oolitic Mn deposits could take place. The deposition of Mn is assumed to be supported by the action of bacteria.

With the oolitic Mn deposits also differences in the nature of the precipitated Mn ores can be noted that depend on the distance from the coast. In nearshore areas first pyrolusite (MnO_2) will form. The pyrolusite zone is followed by a zone with manganite ($Mn_2O_3 \cdot H_2O$) and, remotest from the coast, a zone occurs with rhodochrosite ($MnCO_3$).

The most important deposits of this type are located in the Soviet Union (*Tshiaturi*, to the south of the Caucasus, and *Nicopol* in the Ukraine; Fig. 59).

2.3.2.3. *Organically precipitated deposits* (*deposits of the 'sulphur cycle'*)

These deposits are sedimentary precipitates that can occur both in terrestrial areas of subsidence and in current-less, cut-off, marine basins. Their ore content forms by the interaction between metal-ion-bearing solutions of decomposed rocks, on the one hand, and a H_2S facies, on the other.

The supply of the metal solutions takes place either in connection with the arid concentration deposits of the red-bed type (sulphate, chloride solutions) or under humid climatic conditions in a marine environment as humates, bicarbonates or by means of protective colloids. The elements being supplied are mainly Fe, Cu, Pb, Zn, Ag, Co, Ni, Mo, V and U. The H_2S facies is chiefly produced in an organic 'sulphur cycle' generated by the activity of micro-organisms.

The bacterial *cycle of sulphur* takes place within the environment of formation in two zones, in an upper (aerobic) zone high in oxygen and a lower (anaerobic) zone free of oxygen (Fig. 60):

The formation of H_2S by organisms is possible in the lower, anaerobic range only. H_2S is produced here first by the so-called *desulphurizing bacteria* reducing the inorganic sulphates in the course of their vital functions and, second, to a lesser degree by the *putrefactive bacteria* active on the albumen of dead organisms. As a by-product ammonia is also generated by the so-called ammonia bacteria which thus produce the alkaline solution necessary for the reduction of the sulphates.

In the upper, oxygen-rich zone the H_2S, which diffuses upward from

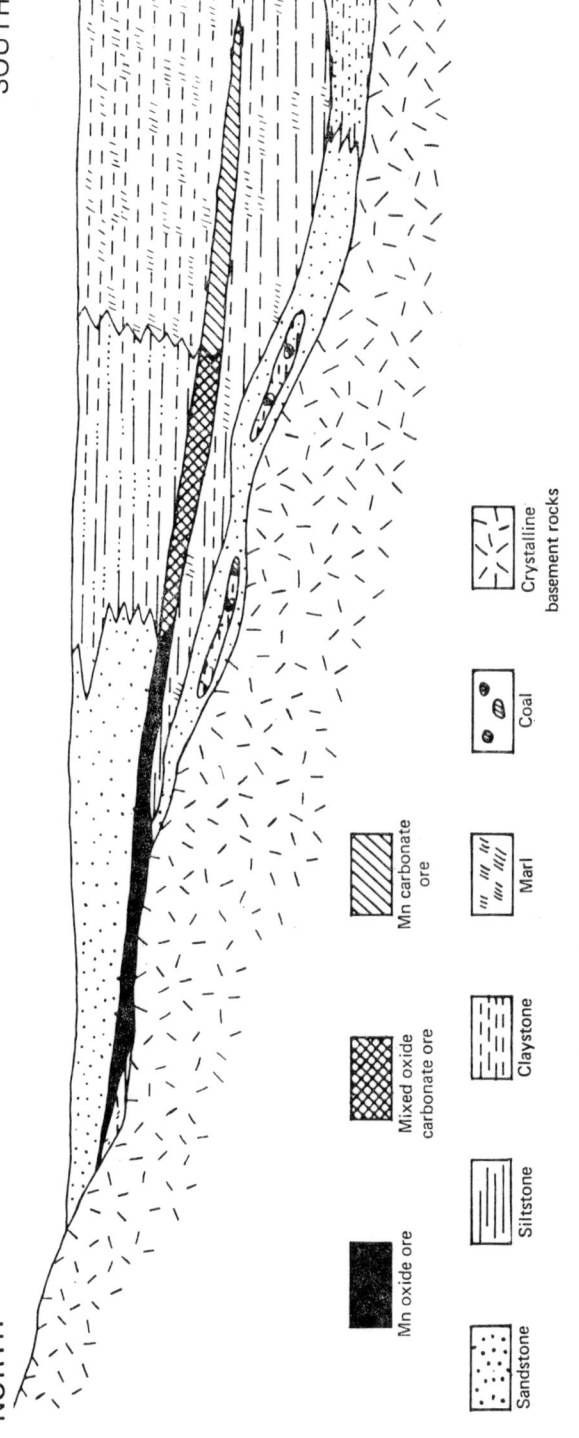

Fig. 59. Diagrammatic cross-section of the Nicopol district, USSR.

the lower, anaerobic zone, is consumed by the *sulphur bacteria* proper. In so doing they oxidize the H_2S to form native S and H_2SO_4. For this they need oxygen; they are aerobic and live therefore in a border-zone which is supplied from above with oxygen and from below with H_2S. In this water level they form a kind of 'bacteria plate'. The H_2SO_4 formed is neutralized by the bicarbonate of Ca, which is always present in surface waters, to yield the poorly soluble $CaSO_4$. The precipitated $CaSO_4$ settles downward into the anaerobic zone and can there be treated again by the desulphurizing bacteria to form H_2S. Thus the sulphur cycle, serving the vital functions of the bacteria, starts afresh.

If this process takes place *undisturbed* only sapropel sediments (i.e. putrefaction products under oxygen-free water) will form at the bottom. These are calcareous-clayey sediments containing considerable amounts of organic matter (bitumen) and some FeS_2 (dark marl slates, alum slate, oil shale). With greater supply of sulphate-bearing solutions, deposits of more or less pure sulphur may occur lying in a marine series (Campagna, Calabria type) and are either pure or mixed with sulphates (gypsum), bituminous marls or clays.

A typical recent example of such a formation occurs in the Black Sea. Here the oxygen-zone goes down to 200 m in the water. Below, down to a depth of about 2000 m, there are water layers without O_2. Abundant animal life can develop only in the oxygen-containing upper layer. Today, at the bottom of the Black Sea, a sapropel facies is forming containing 23-35% of organic matter.

If solutions containing metal ions (solutions of rock decomposition or hydrothermal solutions) reach such areas of deposition they are affected by the presence of H_2S and form metal sulphides. With sufficient supply of cations (Cu, Zn, Ag, Fe, Co, etc), sedimentary sulphide deposits are formed (Fig. 60). The copper content in particular is often so far enriched as to form workable deposits. Further, the bituminous rocks are frequently distinguished by a low content of uranium ($\sim 0.02\%$ U).

Terrestrial deposits of the sulphur cycle may be precipitated in arid basins. In epicontinental seas the formation of marine deposits of the sulphur cycle (copper marl, copper slate—'Kupferschiefer') takes place. Generally, the average metal content of such deposits are not very high. However, these deposits mostly possess a very great areal extension (in contrast to the magmatogene submarine-sedimentary sulphide deposits). They are true sediments and therefore always occur conformably within the stratigraphic complex. The metallization proper is expressed in the form of fine impregnations or lenses. The metal content is very different in the various parts of the areas of deposition. It depends, apart from the supply of the metal, on the environmental conditions (Eh, pH, concentration etc.) and the morphology of the underlying strata.

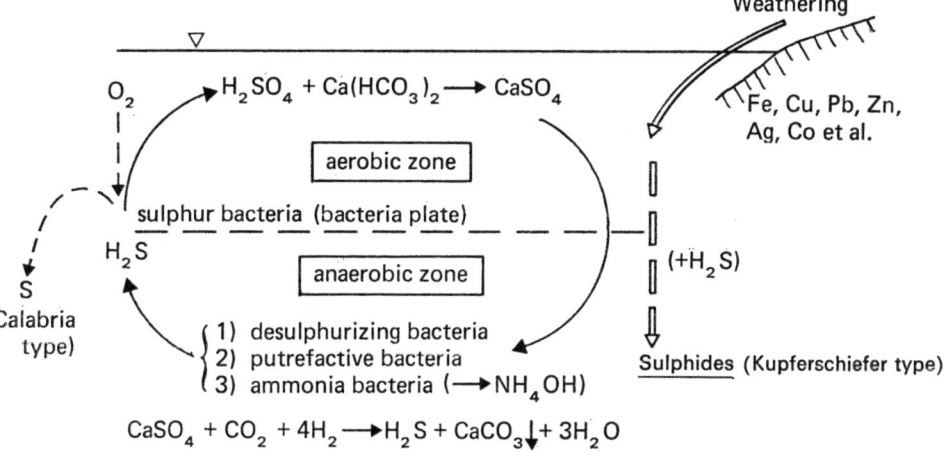

FIG. 60. The 'sulphur cycle'.

(*a*) *Terrestrial precipitated deposits*

Terrestrial precipitated deposits develop from red-bed deposits if the masses of detritus within the arid basins are covered by shallow, mostly stagnant water. It should be noticed, that a distinct separation of the ecological zones of the sulphur bacteria and the desulphurizing bacteria does not always occur here This means an interdigitation of aerobic and anaerobic conditions. Since the unconsolidated nature of the rocks prevents deposition of any massive sulphides, dispersed metallization mainly due to impregnation will take place.

Two important examples of this type of origin are those of *Maubach-Mechernich*, Eifel (Pb, Zn) and the *Katanga-Zambia* copperbelt (Cu, Co).

In the upper Bunter (Triassic Red Sandstone), near *Maubach-Mechernich*, large concentrations of Pb and Zn were formed under arid climatic conditions. Three lens-shaped ore-bodies are situated within a conglomerate-sandstone series 20 m thick. They consist mainly of so-called 'knottenerz' (nodular galena, concretions up to 5 mm in size, partly crusting the quartz grains). The ore minerals (PbS, less ZnS, FeS_2 and $CuFeS_2$) were sometimes redeposited several times. Genetically, we have here an arid basin (red-bed facies) in which primary enriched metals were redeposited by a later H_2S facies and concentrated as sulphides. The overlying parts of the Bunter sandstones have an intensive red colour and do not contain any ore. The contents in the orebodies are up to 3% Pb and 2% Zn. Further, deposits of a similar nature are located near *Freihung* in Oberpfalz (in Keuper sandstone) and *Laisvall*, in North Sweden (in silicified Cambrian sandstones).

The largest copper district of the Earth, *Katanga-Zambia*, also belongs to precipitated deposits of the sulphur cycle. Several ore horizons are embedded in a thick Precambrian (Algonkian) sequence (the 'Katanga system' or 'Roan series'). The rock sequence is made up of conglomerates, sandstones, sandy shales (in part bituminous) and dolomites. The belt extends about 350 km. In Katanga, the ore horizons are chiefly confined to a dolomitic rock facies. The average content of the ore as mined is about 4% copper (chalcopyrite, bornite, chalcocite). The ore reserves are estimated at greater than 500 million tons. Frequently the ores are strongly oxidized (partly thick fossil oxidation and cementation zones containing up to 12% Cu, especially in Katanga). The most important mines are the 'Roan Antelope Mine' in Zambia and the 'Etoile du Congo' in Katanga. The whole district produces annually about 650,000 tons of copper (17% of the world production). In addition to copper further important metal components occur such as Co (about 1% Co as linneite, Co_3S_4) and U in pitchblende. Due to later metamorphism the whole metallization was partly mobilized and more or less redeposited giving hydrothermal replacement structures, partly associated with faults.

Another similar deposit is that of *White Pine*, Michigan.

(b) Marine precipitated deposits

These are sapropelic epicontinental-sea sediments also formed under largely arid climatic conditions with a supply of fresh water containing a high metal content. They are mostly found at the beginning of an evaporite cycle. For deposits to form in this facies the metal should have been preliminarily enriched, for example, in a red-bed facies. The best known deposits of this origin are the *Mansfeld Kupferschiefer* (*copper slate*), G.D.R., and the *copper marl* of Lower Silesia, Poland. The lower wall of the Mansfeld Kupferschiefer is composed of conglomerates (Permian) and sandstone with calcareous-dolomitic intercalations here and there (a typical Permian red-bed facies of the Upper Rotliegendes or lowest Zechstein, respectively). These bottom parts are mostly bleached (indicating a reducing environment and known as 'Weissliegendes') and are in part metalized (the 'sand ore'). Due to a change in facies (from an arid basin to an epicontinental sea) the formation of 'copper slate' took place in the overlying strata. This is a black, bituminous 'marl slate' (50% quartz +sericite, 30% lime + dolomite, 10% bitumen, 10% sulphides) which contains most finely disseminated sulphides (so-called 'speise': chalcopyrite, bornite, chalcosite, sphalerite, less frequently pyrite, galena, tetrahedrite), which represent the maxima of metal deposition. Different layers may be distinguished based upon colour, structures and composition. The hanging wall in this case is composed of grey limestones, which, finally, change into the Zechstein limestone, proper. From the bottom to the top, the bitumen and metal contents decrease whereas the limestone portion increases. As accompanying constituents there are also Mo, V, Ag, Au, Pt, Se and Rh.

As regards regional distribution, distinctly marked and regular zoning can be observed for the main metals. This is largely caused by the so-called '*Rote Fäule*'. Generally the 'Rote Fäule' facies is bound up with rises or sills (such as basement rises, sand bars) within the Zechstein sea. Thus the 'Rote Fäule' is to be regarded during the Lower Zechstein as a facies which formed where the shallow water was well aerated. It should be noted in this, that the 'Rote Fäule-facies' is not confined to a definite horizon, but may run across the time-stratigraphic horizons.

Now, the palaeogeographic situation is not only of importance for the *petrographic features* but also for the *geochemical facies*. Among the geochemical factors of decisive importance for the metal and mineral content are the Eh- and pH-values as well as the solubility products of the sulphides. In the range of the 'Rote Fäule' we have a high redox potential (Eh$>$0). The Fe^{3+}, present here is fixed as hematite, whereas the ions of the nonferrous metals remain in solution and migrate away. In this 'Rote Fäule'-area copper slate and Zechstein limestone have a red (hematite) or green colour (chlorite and biotite). With increasing distance from the 'Rote Fäule'-area the solutions containing nonferrous metals once more find themselves in a more intensely reducing environment. Mainly copper sulphides (chalcosite-bornite-chalcopyrite) form in the transitional region with a Eh-value $\approx$ 0·0 because of the negligible solubility product. Under increasingly stronger reducing conditions (Eh$<$0) first the solubility product of the PbS and finally that of ZnS is reached (solubility product $Cu_2S = 10^{-50}$, solubility product $PbS = 10^{-28}$, solubility product $ZnS = 10^{-25}$). Thus, moving away from the 'Rote Fäule'-region, we find a copper facies and, at a greater distance, a Pb facies and then a Zn facies. Note that the Kupferschiefer enriched with Cu occurs only in a slightly reducing environment, that is found in the transitional region between the 'Rote Fäule'-facies and the 'sapropel facies'.

Consequently the most important result of these recent investigations is the realisation that the greatest concentrations of Cu in the Kupferschiefer run parallel to the 'Rote Fäule'-areas. Thus mapping-out of the 'Rote Fäule' is today the most significant feature in the exploration of these deposits.

The Kupferschiefer proper is mined in the *Mansfeld* and *Sangerhausen* troughs (G.D.R.) as well as near Richelsdorf, Hessen (F.R.G.). The Cu content in Lower Zechstein rocks, however, extends far beyond these regions although it is not everywhere workable. The extent is controlled by the area of the Zechstein sea, which at that time spread

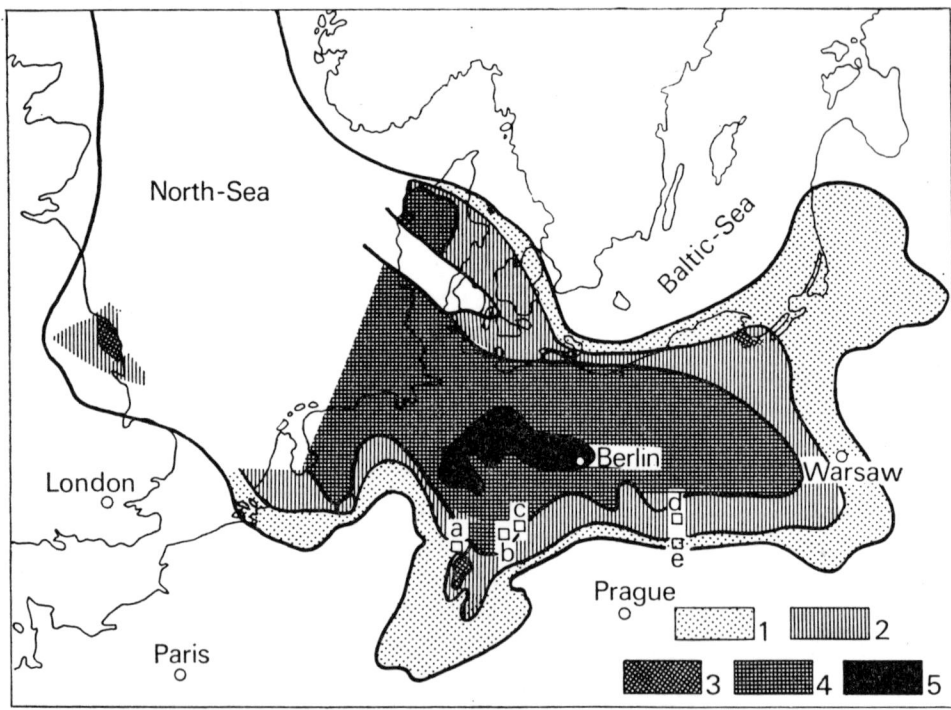

FIG. 61. Extent of the Zechstein sea in Central Europe. The Kupferschiefer occurs at the base of the Upper Permian (Zechstein). In the upperwall are the salt deposits of the Zechstein (Z 1-3) (from Rösler, Baumann and Jung, 1968).
1–marginal zone,
2–distribution of rock salt,
3–distribution of potassium salt Z 1,
4–distribution of potassium salt Z 2,
5–distribution of potassium salt Z 3

Noteworthy mining districts of the Kupferschiefer:
a–Richelsdorfer Mountains,
b–Sangerhausen district,
c–Mansfeld syncline,
d–Presudetic monocline,
e–Innersudetic syncline.

600 km west-east from the edge of the Rhenish Massif to Lower Silesia and 200 km north-south from the Thuringian Forest to the North German Plain. In the near-shore zones (edge of the Sudetic Mountains and the Rhenish Massif) the 'slate' horizon takes the form of calcareous copper marl. In Lower Silesia the copper marl occurring with thicknesses up to 3 m, is workable in the trough of *Zajaczek-Grodzice*, Poland. As regards extension and metal reserves these occurrences surpass the Mansfeld deposits (Fig. 61).

The 'marl slate' in *Durham*, England, is similar to the Kupferschiefer as regards stratigraphy and facies although not of economic importance.

3
Metamorphic transformation of ore deposits

If sedimentary or magmatic ore deposits are buried in *deeper zones* of the earth by geological processes, or if due to *tectonic movements* they happen to be located in pressure and temperature ranges higher than those prevailing during their formation, they will be metamorphosed. In this transformation the *chemical* composition may be only slightly changed. The changes affect chiefly the *mineral composition* and *fabric*. The changes in mineral composition are first of all influenced by temperature, whereas changes in fabric are mainly caused by pressure— especially directed pressure. Physicochemically, metamorphism means the establishment of a new equilibrium between co-existing mineral phases. Thus, in general new concentrations of elements, that is the formation of ore deposits, does not take place in the course of these processes.

From a purely *genetic* viewpoint there is no reason for separating out a special group 'metamorphic deposits'. Each so-called 'metamorphic deposit' has usually been developed from a deposit of the endogenetic or exogenetic cycles. On the other hand, each deposit has been subjected, in the geologic epochs after formation, to more or less metamorphic influences. As far as possible we have described deposits in this account (even though metamorphosed) according to their probable primary mode of origin. Some deposits, however, have been subject to such considerable changes by metamorphism, that their original genesis is very difficult to recognize.

3.1. Causes and factors acting in metamorphism

Temperature and pressure are the principal factors in metamorphism. The deeper the primary rock or the primary deposit is located inside the earth's crust due to superposition or tectonic processes the higher the *temperature*. This burial acts regionally. A temperature rise can also be brought about by a magmatic intrusive body. The resultant contact metamorphism is of more local significance. It is mostly accompanied by gas transfer and impregnation by magmatic solutions.

Pressure, too, increases with increasing depth. Part of the pressure is hydrostatic and depends only on the thickness of the overlying strata. In many regions, especially orogenic areas, directed pressures also occur.

If the resulting strain exceeds the limit of plasticity, fracturing and crushing take place (mylonites, pseudotachylites) followed by recrystallization processes. Interstitial water and the gases and solutions present in the capillary interstices become mobilized under the influence of temperature and pressure. These react with the solid mineral constituents and bring about *isochemical* reactions (i.e. there is no change in the overall composition of the rock mass). Additional magmatic substances may be supplied in deeper zones of the earth resulting in *allochemical* (metasomatic) reactions.

The effects brought about by the factors temperature and pressure are chiefly:

(*a*) modification of existing minerals;
(*b*) growth of new minerals;
(*c*) development of a new fabric;
(*d*) change in shape of the deposit.

3.2. Kinds of metamorphism and its products

Depending on the temperature and pressure conditions we can distinguish four kinds of metamorphism:

(1) *Dynamic-metamorphism* (isochemical; up to plus 300 °C and predominantly shows the effects of deformation).

The rocks formed are the crystalline schists of the so-called epi- to mesozones. Deposits lying within these rock series are subjected to changes of their mineral composition and fabric.

(2) *Thermal metamorphism* (up to 600 °C or over)

Thermal metamorphism is caused by intruding magma bodies. It is mostly static and not influenced by movements. There can be distinguished
 (*a*) without supply of material (only heat action, isochemical);
 (*b*) with supply of material (contact metasomatism, hydrothermal replacement, i.e. allochemical).
The intensive action of heat (with or without the supply of juvenile material) causes such rocks as quartzite, hornfels and marble to form. The pre-existing deposits present in the aureole zone are thermally altered.

(3) *Injection metamorphism and migmatization* (allochemical; > 600 °C)

Here, magmatic supplies play a greater role. They penetrate the country rock in many ways and cause gaseous transfer. Injection gneisses and remelted rocks form. Earlier deposits may change their metal associations owing to fluid supplies, and very uncommon element combinations may occur.

(4) *Retrograde metamorphism*

If already metamorphosed deposits or deep-seated magmatic occurrences come to be situated in higher zones within the earth in the course of geologic development they are once more subjected to changes, giving rise to 'retrograde' or regressive metamorphism, the so-called diaphthoresis. This process involves for example, the Fe-minerals (garnet, staurolite, biotite) being transformed to chlorite, that is, to Mg-minerals. In this transformation a great amount of the iron is liberated; it may be precipitated again at other suitable places and concentrated to a certain degree. Some workers suggest such an origin for the metasomatic deposits of siderite and magnesite in the eastern Alps. At any rate diaphthoretic mobilization of metal solutions is one of the important working hypotheses for the formation of ore deposits by metamorphism.

3.3. The formation of metamorphic ore deposits

The peculiarity of ore metamorphism is, that the ore minerals generally are much more sensitive than the rock-forming minerals as regards the operation of metamorphism. Thus many ore deposits reveal metamorphic influences when the enclosing rocks reveal no or only very slight indications of such changes. At relatively small depths the temperature is already about 100 °C, and water is always present as interstitial water. These correspond to the epithermal conditions of the formation of hydrothermal ore deposits. Therefore, for example, a sedimentary deposit originated under surface conditions may be transformed metamorphically in this zone. The overall appearance of this deposit, which formed as a sedimentary deposit, has become 'hydrothermal'. For these reasons many primarily sedimentary deposits have been interpreted as hydrothermal deposits, and their genesis has given rise to much discussion. Such deposits, however, were formed as purely sedimentary deposits and changed only later under 'hydrothermal' conditions by slight metamorphism. Magmatic influences need not be assumed to have had any effect either in transformation or in the primary deposition.

Each process of metamorphism, including its retrograde development, may result in pegmatitic-pneumatolytic and hydrothermal sequences corresponding to the pressure-temperature conditions and analogous to the magmatic cycle. Such convergence renders the recognition of the primary origin of a given deposit very difficult.

A simple classification of metamorphic ore deposits is hardly possible, the source material and the metamorphic processes being so very manifold. An expedient division is that based upon the predominant kind of metamorphism:

(a) static (load metamorphism or general metamorphism: mainly in geosynclines);
(b) kinetic (movement metamorphism or dynamic metamorphism: mainly in the orogenic);
(c) thermal (contact metamorphism: by intrusive rocks).

3.3.1. Transformation of deposits by static-kinetic metamorphism

To this group belong deposits that were subjected to distinct changes in the *mineral composition*, the *fabric*, and the geologic *shape* due to regional metamorphism or dynamometamorphism (statically or kinetically). In the transformation, both higher pressures, especially directed pressures, and higher temperatures acted. Transformations of this kind take place first of all in the range of the epi- and meso-zones. In general, a supply of substances from outside did not occur during recrystallization.

Hydrothermal recrystallization occurs very frequently and affects ore deposits formed at low temperatures and pressures (sedimentary and magmatic-sedimentary deposits). Primary salt-bearing water is always included in sediments in the form of pore water. Consequently with burial at relatively shallow depths and under the prevailing temperature conditions a solution may already form that reveals a fully 'hydrothermal' nature. It then brings about the recrystallization of especially sensitive minerals (above all sulphides, organic matter, salts).

Sometimes the secondary hydrothermal solutions may be transported into higher 'stockworks' with a younger rock facies and separate out there as 'regenerated deposits'.

The new *mineral composition* depends principally on the source composition, the degree of metamorphism and the country rock, with which certain constituents may be exchanged (hydrated ores are dehydrated to form oxides: limonite $\longrightarrow$ hematite or magnetite; bauxite $\longrightarrow$ corundum).

The *fabric* becomes much more altered since most ore minerals are very sensitive to kinetic stress. The behaviour of the ore minerals is very variable. The grains of the most brittle ore minerals are broken (for example pyrite, arsenopyrite). Some are not as brittle (sphalerite, pyrrhotite) and may undergo plastic deformation, giving rise to gneiss-like structures. Subsequently, recrystallization or new crystallization take place at greater depths. With predominant hydrostatic pressure the structure is granular, with directed pressure it is schistose.

Finally, the *shape* of the ore deposits may be largely changed by metamorphism.

Geologically, the deposits transformed by regional metamorphism must occur within metamorphic rocks. Thus they are preferentially confined to the ancient, consolidated (mostly Precambrian) shields. Well-known examples of such metamorphically transformed deposits are the so-called '*itabiritic*' iron-ore deposits of Lake Superior, USA, of Krivoi Rog, Ukraine, the banded deposits of *manganese* ore of Postmasburg in South Africa, the *sulphide* deposits of the Rammelsberg, Harz Mountains, and Falun, Sweden.

80% of the total iron production of the USA up to 1950 came from the iron-ore deposits of *Lake Superior*. Within the Archean (Keewatin) rocks, fine-grained, banded hematite-quartzites (jaspilites, taconite) are found partly associated with 'greenstones'. In addition to hematite and magnetite, siderite and iron silicates (greenalite) also occur. These 'taconite ores' ($\sim 30\%$ Fe $+50\%$ SiO_2) amounting to about 70,000 million tons are the great ore reserve of this region. They represent probably a metamorphic product of Lahn-Dill type of deposit. The ores mined up till now have been mainly primary taconite ores later re-deposited after a period of weathering in Precambrian times and considerably enriched by the removal of SiO_2 ($\sim 50\%$ Fe). The enriched ores were metamorphosed in the Precambrian. The chief mining district has been the *Mesabi Range* with the largest ore open-cuts of the world and an annual output of crude ore up to 70 million tons (Fig. 62).

A similar deposit is that of *Krivoi Rog* in the Ukraine, which consists of banded iron-bearing quartzites up to 50 m thick, within the Krivoi-Rog series of early Algonkian age (Fig. 63). The deposit of *Kursk* also belongs to the same formation. It had been well known for a long time due to a strong magnetic anomaly and has been exploited from under Mesozoic overburden up to 300 m thick. Further giant occurrences are situated at *Minas Gerais*, Brazil, and in the *Singhbhum district*, India (which produces about 50% of India's total iron production).

Probably of similar origin are the commercially very important Mn deposits near *Postmasburg*, South Africa, (in Paleozoic rocks: banded Mn-silicate deposits with intercalated hematite and hornstone), in *Ghana* (Mn-garnetiferous schist with green schists) and near *Nagpur*, India, (banded Mn-silicates with Mn-garnet and quartzite: the so-called gondites). These extensive occurrences are especially workable where they have undergone repeated enrichment by weathering (from about 25% to about 45% Mn).

A similar occurrence is the polymetallic sulphide deposit of *Rammelsberg* near Goslar, Harz Mountains (F.R.G.). It is one of the most valuable and important deposits of non-ferrous metals in West Germany and has been mined since A.D. 900. In the strongly folded Middle-Devonian schists of Wissenbach (predominantly black shales) two thick lens-shaped ore bodies are embedded. The whole forms part of the Oberharz Devonian anticline on whose northwestern flank intensive folding occurs and the ore-bodies are confined to a syncline (Fig. 64). The 'Altes Lager' ('Old Deposit', 500 m × 300 m in area; thickness up to 30 m) is almost exhausted. The 'Neues Lager' ('New Deposit') is developed down dip for a distance of 600 m (thickness up to 50 m). The Rammelsberg ores are principally a most fine-grained mixture of pyrite, chalcopyrite, sphalerite, galena and baryte (30% FeS_2, 19% Zn, 9% Pb, 1% Cu, 25% $BaSO_4$). The ore of the two main deposits is most impressive for its fine banding and its intensive folding. As regards the ore distribution it is of interest to note that pyrite and chalcopyrite occur chiefly in the lower parts of the ore

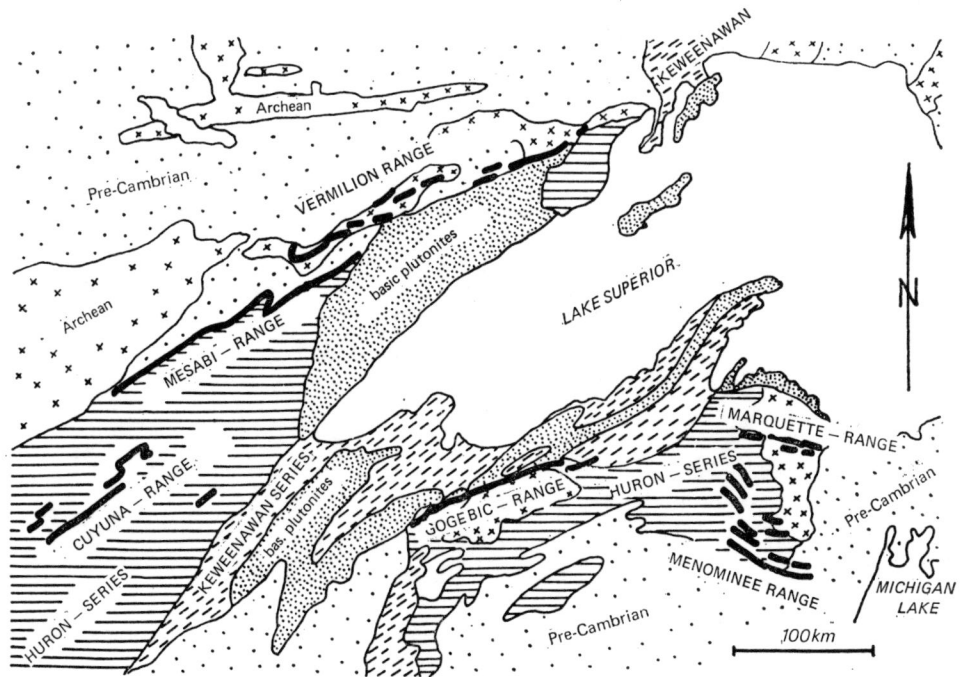

FIG. 62. Iron-ore deposits, region of Lake Superior, USA and Canada (from Zeschke, 1964).

W
E
Krivoi-Rog
Syncline
1 – Plagiogranites
2 – Amphibolites
3 – Lower Krivoi-Rog series (amphibole schists, quartzites, mica schists and garnet schists)
4 – Carbonate-talc horizon
5 – Central productive series with 7 iron ore horizons (ores of magnetite, martite, hematite; jaspilite)
6 – Schists of central series (chlorite-biotite schists and amphibole schists)
7 – Upper Krivoi-Rog series (dolomites, sandstones, micaceous and amphibole schists)
8 – Faults
9 – Younger microcline-plagioclase granites

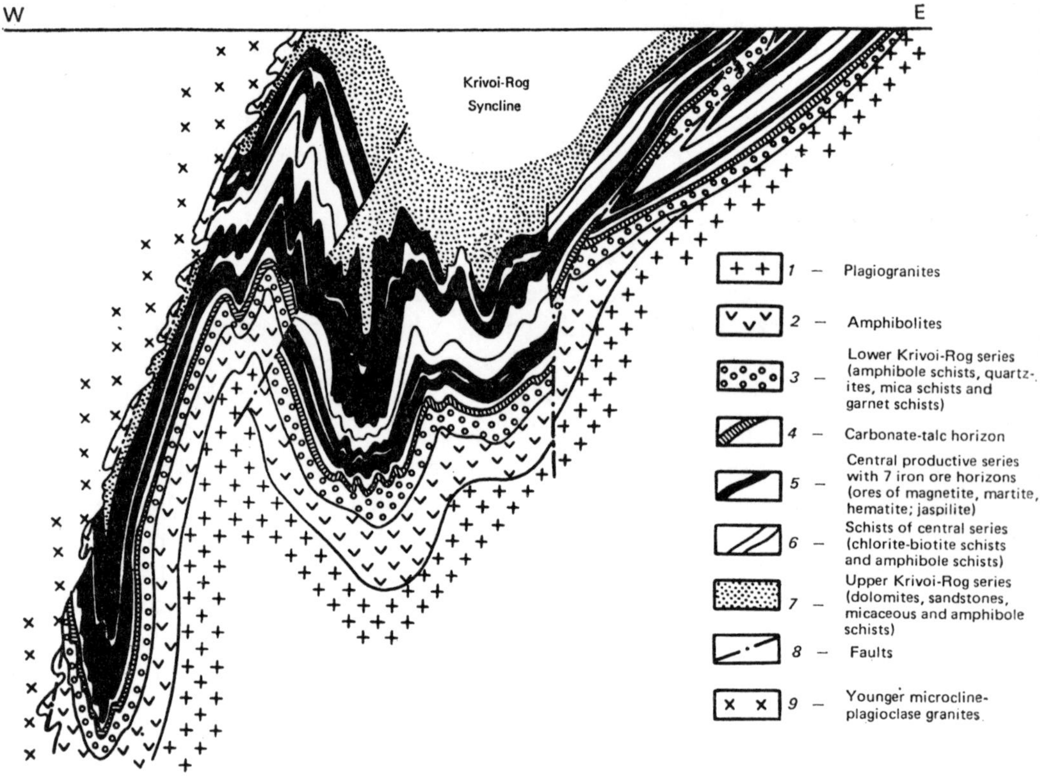

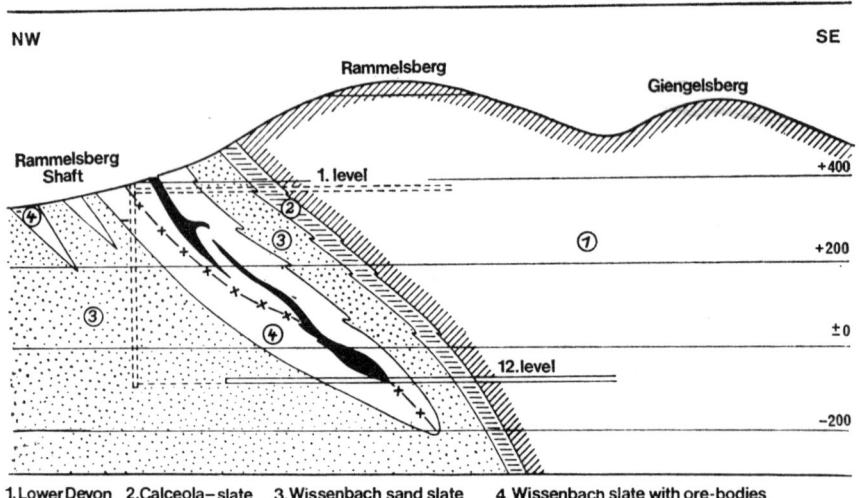

FIG. 63. Geological cross-section of Krivoi-Rog (from V. I. Smirnov, 1970).

NW
SE
Rammelsberg
Giengelsberg
Rammelsberg Shaft
1. level
12. level
+400
+200
±0
-200
1. Lower Devon 2. Calceola-slate 3. Wissenbach sand slate 4. Wissenbach slate with ore-bodies

FIG. 64. Cross-section through the Rammelsberg deposit (after Kraume, 1960).

deposits, whereas galena and baryte concentrate in the upper parts of the deposit.

Similar origin and composition is shown by a number of the '*metamorphic sulphide beds or deposits*', especially in Scandinavia (*Norway*: Leksdal, Röros, Lökken, Sulitjelma; *Sweden*: Falun, Boliden/Skellefte district, within the leptite formation, with great amounts of arsenopyrite and gold; *Finland*: Outukumpu, numbers among the largest copper deposits of Europe) and in the basement of *Canada* (Rouyn-Noranda/Quebec, Flin-Flon/Manitoba, and others).

3.3.2. Transformation of deposits by contact metamorphism

We use the term contact metamorphism for all mineral and chemical changes as well as changes of fabric, that are brought about by the purely thermal action of a rising magma on its environment. Therefore the action of contact metamorphism is confined to a zone immediately surrounding the intrusive body and decreases gradually outward. The contact action may be exogene (in the country rock) and endogene (in the igneous rock). Further, contact metamorphism may take place isochemically or allochemically. In general no supply of substance takes place here from the magma (only volatiles). If any important movement of material takes place, especially of heavy metals, contact-pneumatolytic or metasomatic deposits respectively, will form, which belong to the magmatic occurrences. These may be recognized by the fact that within the contact zone ore minerals occur, which do not occur in the rock outside it.

It is especially the *mineral composition* that is changed by contact metamorphism. Water-bearing oxides are changed to the corresponding oxides free of water (for example limonite and siderite to magnetite; pyrolusite and psilomelane to braunite and hausmannite; chamosite to Fe-olivine; bauxite to corundum; pyrite to pyrrhotite; carbonaceous rocks to graphite). As to *fabric*, mostly a coarsening of the grain takes place. The *shape* of the rock-bodies affected by the metamorphism is usually not changed.

Deposits, produced purely by contact-metamorphism are relatively rare (corundum deposit of *Smyrna*, vanadium deposit of *Mina Ragra/Peru*). Mina Ragra/Peru is the richest vanadium deposit (5000 m above sea level and one of the highest mines in the world). Bituminous marly shales (partly containing asphalt) of the Eocene were contact-metamorphised by porphyry veins. In this, patronite and an asphalt coke high in vanadium were formed (up to 13% V). This deposit yields the greater part of the world production (more than 100 tons per year of vanadium).

3.3.3. Polymetamorphic ore deposits

These complex metamorphic deposits originate in the deepest metamorphic zones of the earth's crust. The transformations take place here under physico-chemical conditions that are very similar to magmatic conditions. Thus, injections of magmatic melts occur, and

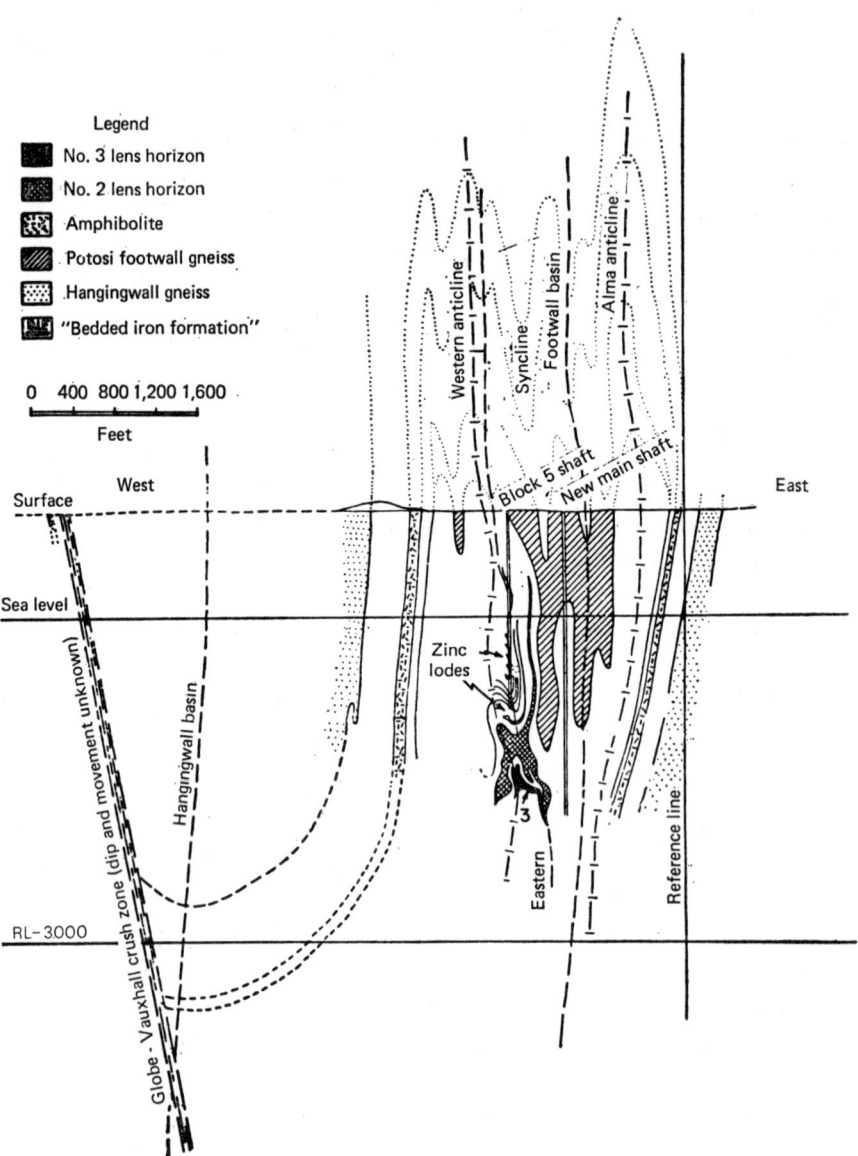

FIG. 65. Cross-section, Broken Hill deposit, Australia (from Park, Jr. and MacDiarmid, 1964).

pegmatitic-pneumatolytic solutions can penetrate into the rocks and deposits buried in these zones. Due to this action dissolution and migration of the material making up the deposits may occur. The mobilized substances then mix with primary supplies and separate out in other places. Places preferred for reprecipitation are mainly limestones and dolomites. The formation of new deposits with an extraordinarily complex *mineral composition* can take place. The presence of very rare and unique metal associations and assemblages always indicates the possibility of metamorphic mobilization. The mobilized solutions mostly contain substances from different ore occurrences and rocks that, if precipitated at the same place, may give rise to most uncommon mineral assemblages. Such deposits are called polymetamorphic because the most different metamorphic factors participated in their formation.

In general the *shape* of these deposits is very variable. Some of these deposits are situated today, in the form of lenses, in metamorphic rocks (gneisses, leptites, calc-silicate rocks). But in the form of irregular bodies they are also largely bound to metamorphosed limestones and dolomites and resemble in their geological features contact-metasomatic deposits.

Polymetamorphic deposits occur mainly in the old Pre-cambrian shields of the earth and are always confined to highly metamorphic rocks. The Pb-Zn deposit of *Broken Hill*, Australia, is a deposit that has been strongly transformed by polymetamorphism. For output and reserves it is at present the largest Pb-Zn deposit in the world. The primary deposit was certainly layered (submarine layers?). The ore bodies, up to 45 m thick, lie conformably in steeply folded Precambrian schists (Fig. 65). They are composed of manifold sulphides (e.g. PbS, ZnS) as well as Mn-garnet and Zn-spinel. The annual output is about 1·5 million tons with contents of 14% Pb and 14% Zn(!).

A further example is *Franklin*, New Jersey, from which are derived franklinite, $(Zn, Mn)(Fe, Mn)O_4$, zincite, ZnO, and another 150 very uncommon minerals (contents: 17% Zn, 10% Mn). The deposit is presumably the result of high-grade changes induced by contact metamorphism and metasomatism because the ore-bodies occur in marble in contact with an injection gneiss.

Bibliography

Barnes, H. L. and G. Kullerud (1961): Equilibria in sulfur-containing aqueous solutions, in the system Fe-S-O, and their correlation during ore deposition. *Econ. Geol.*, **56**, pp. 648-688.
Barsukov, V. L. (1957): Zur Geochemie des Zinns. *Geochimija*, H. 1, 36-45 (Russ.).
Barton, P. (1959): The chemical environment of ore deposition and the problem of low-temperature ore transport. In: P. H. Abelson (Editor), *Researches in Geochemistry*, pp. 279-300. New York.
Bateman, A. M. (1951): *Economic Mineral Deposits*. New York, John Wiley & Sons, Inc., second edition.
Bateman, A. M. (1964): *The Formation of Mineral Deposits*. New York and London, J. Wiley & Sons, Inc.
Baumann, L. (1958): Tektonik und Genesis der Erzlagerstätte von Freiberg. *Freib. Forsch. H.* C 46, pp. 208.
Baumann, L. (1965): On the zonal distribution of mineralization in the veins of the Freiberg ore district. 1. IAGOD-Symposium, Prague. *Problems of postmagmatic ore deposition*, Vol. II, pp. 56-66.
Baumann, L. (1965): Die Erzlagerstätten der Freiberger Randgebiete. *Freib. Forsch. H.* C. 188.
Baumann, L. (1967): Zur Frage der varistische nund postvaristischen Mineralisation im sächsischen Erzgebirge. *Freib. Forsch. H.* C pp. 15- 38.
Baumann, L. (1970): Tin deposits of the Erzgebirge. *Transact. Section B, Inst. of Mining and Metallurgy*, Vol. 79, pp. 68-75.
Bentz, A. and H. J. Martini (1968): *Lehrbuch der angewandten Geologie*, 2. Bd., Teil 1: *Geowissenschaftliche Methoden*. Ferdinand Enke Verl., Stuttgart.
Bolduan, H. (1963): Geologie und Genese der Zinn-Wolframlagerstätte Geyer (Erzgeb.). *Freib. Forsch. H.* C, pp. 7-34.
Borchert, H. (1960): Geosynklinale Lagerstätten, was dazu gehört und was nicht dazu gehört, sowie deren Beziehungen zu Geotektonik und Magmatismus. *Freib. Forsch. H.* C 79, pp. 7-61.
Borchert, H. (1964): Über Faziestypen von marinen Eisenerzlagerstätten. *Ber. Geol. Ges.*, Bd. 9, H. 2, pp. 163-193.
Butler, B. S. (1929): Relation of the ore deposits of the southern Rocky Mountains region to the Colorado Plateau. *Colo. Scient. Soc. Proc.*, **12**, Nr. 2. Denver, Colo., pp. 23-36.
Cloos, H. (1936): *Einführung in die Geologie*. Berlin.
Cotta, B. v. (1864): *Erzlagerstätten im Banat und in Serbien*. Wien: Braumüller.
Ehrenberg, H., A. Pilger and F. Schröder (1954): Das Schwefelkies-Zink-

blende-Schwerspatlager von Meggen (Westfalen). *Monographien d. Deutschen Blei-Zink-Lagerstätten*, 7. Beih. z. *Geolog. Jahrb.*, H. 12.

Finch, W. I. (1967): Geology of epigenetic uranium deposits in sandstone in the United States. *Geological Surv. Prof. Paper* **538**, Washington.

Goldschmidt, V. M. (1938): Geochemische Verteilungsgesetze der Elemente. *I-IX. Akad. Wiss. Oslo, math.-naturw. Kl.* 1923-1927.

Goldschmidt, V. M. (1954): *Geochemistry*. Oxford, Clarendon Press.

Goranson, R. W. (1936): Silicate-water systems: The solubility of water in Albite melt. *Trans. Amer. Geophysic. Union*, **17**.

Hosking, K. F. G. (1965): The Search for Tin. *Mining Magazine*, **113**, pp. 261-273, 308-383, 448-461

Kraume, E. (1960): Stratigraphie und Tektonik der Rammelsberger Erzlager unter besonderer Berücksichtigung des Neuen Lagers unter der 10. Sohle. *Erzmetall*, Bd. XIII, H. 1, pp. 7-12.

Krauskopf, K. B. (1957): The heavy metal content of magmatic vapour at 600 °C. *Econ. Geol.*, **52**, pp. 786-807.

Kauskopf, K. B. (1964): The possible role of volatile metal compounds in ore genesis. *Econ Geol.*, **59**, pp. 22-45.

Mason, B. (1952): *Principles of geochemistry*. J. Wiley & Sons, New York.

Mertie, J. B. Jr. (1969): Economic Geology of the Platinum Metals. *Geological Survey Prof. Paper* **630**, Washington.

Metz, K. (1967): *Lehrbuch der tektonischen Geologie*. 2. Aufl. Ferdinand Enke Verl., Stuttgart.

Naldrett, A. J. and G. Kullerud (1967): A study of the Strathcona Mine and its Bearing on the Origin of the Nickel-Copper Ores of the Sudbury District, Ontario. *Journal of Petrology*, **8**, No. 3, pp. 453-531.

Niggli, P. (1952): *Gesteine und Minerallagerstätten*. Verl. Birkhäuser, Basel.

Oelsner, O. (1958): Die erzgebirgischen Granite, ihre Vererzung und die Stellung der Bi-Co-Ni-Formation innerhalb dieser Vererzung. *Geologie*, 7. Jg., H. 3-6, pp. 682-701.

Park, Ch. F. Jr. and R. A. MacDiarmid (1964): *Ore Deposits*. W. H. Freeman and Comp., San Francisco and London.

Petrascheck, W. E. (1961): *Lagerstättenlehre*. Springer Verl., Wien.

Ridge, J. D. (1968): *Ore Deposits of the United States*, 1933-1967, Vols. I and II, Am. Inst. of Min., Met. and Petr. Eng., Inc., New York.

Rösler, H. J., L. Baumann and W. Jung (1968): Postmagmatic Mineral Deposits of the Northern Edge of the Bohemian Massif (Erzgebirge-Harz). *International Geological Congress*, XXII Session, Prague, 57 (Guide to Excursion 22 AC (c) East Germany.)

Rösler, H. J. and L. Baumann (1970): On the different Origin of Variscan and Post-Variscan ('Saxonic') Mineralizations in Central Europe. *Intern. Union Geol. Sci. A, No. 2: Problems of Hydrothermal Ore Deposition*, Schweizerbart Verl., Stuttgart, pp. 71-77.

Schneiderhöhn, H. (1941): *Lehrbuch der Lagerstättenkunde*. Erster Bd. Verl. Gustav Fischer, Jena.

Schneiderhöhn, H. (1962): *Erzlagerstätten.* Gustav Fischer Verl., Stuttgart.
Schneiderhöhn, H. (1958): *Die Erzlagerstätten der Erde.* Bd. I (*Die Erzlagerstätten der Frühkristallisation*). Gustav Fischer Verl., Stuttgart.
Schneiderhöhn, H. (1961): *Die Erzlagerstätten der Erde.* Bd. II (*Die Pegmatite*). Gustav Fischer Verl., Stuttgart.
Shearman, D. J. (1966): Origin of marine evaporites by diagenesis. *Trans. Inst. Mining Met.*, B 75, pp. 207-215.
Skácel, J. (1966): Die Eisenerzlagerstätten des mährisch-schlesischen Devons. *Rozpravy, řada, ročnik,* **76,** 11, pp. 59.
Smirnov, V. I. (1968): The Sources of Ore-Forming Material. *Econ. Geol.,* **63,** pp. 380-389.
Smirnov, V. I. (1970): *Geologie der Lagerstätten mineralischer Rohstoffe.* VEB Deutscher Verl. f. Grundstoffindustrie, Leipzig.
Stille, H. (1940): Zur Herkunft der Magmen (Source of magma). *Abh. Preuss. Akad. d. Wiss., Math.-Nat. Kl.,* No. 19, p. 31.
Varentsov, I. M. (1964): *Sedimentary manganese ores.* Elsevier Publishing Co., Amsterdam.
Verhoogen, J. (1938): Thermodynamical calculation of the solubility of some important sulphides, up to 400 °C. *Econ. Geol.* **13,** pp. 34-51.
Vogt, J. H. L. (1893): Bildung von Erzlagerstätten durch Differentiationsprozesse in basischen Eruptivgesteinsmassen. *Zeitschr. f. prakt. Geologie,* 4-11, pp. 257-284.
Williams, D. (1934): The geology of the Rio Tinto mines, Spain. *Trans. Inst. Min. Metall., London,* **43,** pp. 593-678.
Williams, D. (1961): Further reflections on the origin of the porphyries and ores of Rio Tinto, Spain. *Trans. Inst. Min. Metall., London,* **71,** 265-266.
Williams, D. (1965): Vulkanismus und Erzlagerstätten. *Bergakademie,* 17. Jg., H. 10, pp. 591-599.
Zeschke, G. (1964): *Prospektion und feldmässige Beurteilung von Lagerstätten.* Springer Verl., Wien.

Appendix

Metal-content of the most important ore-minerals, arranged according to the Periodic System.
(after Schneiderhöhn, 1962).

Atom. Nr.	Element		Mineral	Formula	Metal-content in %		Density
					theoretical	actual	
3	Lithium	Li	Spodumene	LiAl[Si$_2$O$_6$]	3.73 Li	1.34–3.43 Li	3.1–3.2
			Amblygonite	LiAl[F, OH/PO$_4$]	4.7 Li	3.3–4.67 Li	3
			Triphylite	Li(Fe, Mn) [PO$_4$]	4.4 Li		3.5
			Lithiophilite				
			Lepidolite	KLi$_2$Al[(F, OH)$_2$/Si$_4$O$_{10}$] (?) KLi$_{1.5}$Al$_{1.5}$ [(F, OH, $^1/_2$O)$_2$/AlSi$_3$O$_{10}$]	3.56–3.69 Li to 2.61–2.76 Li	0.6–2.76 Li	2.8–2.9
			Zinnwaldite	KLiFeAl[(F, OH)$_2$/AlSi$_3$O$_{10}$]		1.58–1.60 Li	2.9–3.1
			Petalite	Li[AlSi$_4$O$_{10}$]	2.2 Li		2.4
11	Sodium	Na	Halite	NaCl	39.3 Na		2.2
19	Potassium	K	Sylvite	KCl	52.4 K		2
			Carnallite	KCl · MgCl$_2$ · 6 H$_2$O	14.1 K		1.6
			Leucite	K[AlSi$_2$O$_6$]	18 K		2.45
			Kainite	K, Mg[Cl/SO$_4$] · 3 H$_2$O	15.7 K		2.1
37	Rubidium	Rb	Carnallite	see above		0.015–0.037 Rb	
			Lepidolite	see above		1.19–3.46 Rb	
53	Cesium	Cs	Lepidolite	see above		0.075–0.68 Cs	
			Pollucite	(Cs, Na) [AlSi$_2$O$_6$] · H$_2$O <1	42.8 Cs		2.9
4	Beryllium	Be	Beryl	Al$_2$Be$_3$ [Si$_6$O$_{18}$]	5.07 Be	2.8–5.4 Be	2.7
			Chrysoberyl	Al$_2$BeO$_4$	7.15 Be		3.7
			Helvite	(Mn, Fe, Zn)$_8$[S$_2$/(BeSiO$_4$)$_6$]			3.1–3.4
12	Magnesium	Mg	Magnesite	MgCO$_3$	28.8 Mg		3.0
			Dolomite	CaMg [CO$_3$]$_2$	13.15 Mg		2.9
			Carnallite	see above	8.74 Mg		1.6
			Kieserite	MgSO$_4$ · H$_2$O	17.6 Mg		2.57
			Olivine	(Mg, Fe)$_2$ [SiO$_4$]	34.4 Mg		3.3

APPENDIX

Atom. Nr.	Element		Mineral	Formula	Metal-content in %		Density
					theoretical	actual	
20	Calcium	Ca	Calcite	$CaCO_3$	40 Ca		2.7
			Anhydrite	$CaSO_4$	29.4 Ca		3.0
			Gypsum	$CaSO_4 \cdot 2 H_2O$	23.3 Ca		2.3
38	Strontium	Sr	Strontianite	$SrCO_3$	59.3 Sr		3.7
			Celestite	$SrSO_4$	47.7 Sr		3.9
56	Barium	Ba	Barite	$BaSO_4$	58 Ba		4.3
			Witherite	$BaCO_3$	69.5 Ba		4.5
13	Aluminum	Al	Hydrargillite (Gibbsite)	$Al(OH)_3$	34.7 Al		2.3
			Diaspore	$AlOOH$	45.0 Al		3.4
			Cryolite	$Na_3[AlF_6]$	12.8 Al		2.9
			Kaolinite	$Al_4[(OH)_8/Si_4O_{10}]$	20.9 Al		2.6
			[Nepheline]	$KNa_3[AlSiO_4]_4$			
39	Yttrium	Y	Gadolinite	$Y_2FeBe_2[O/SiO_4]_2$	38 Y	1.6–11.7 Y	4.5
			Samarskite	$(Y, Er)_4[(Nb, Ta)_2O_7]_3$	20–28.4 Y		6
57 71	Cerium etc.	Ce	Monazite	$Ce[PO_4]$	59.7 Ce	17.1–30.3 Ce	5
			Orthite (Allanite)	$(Ca, Ce, La, Na)_2)Al, Fe, Be, Mg, Mn)_3[O/OH/SiO_4/Si_2O_7]$			
			Pyrochlor (Koppite)	$(Na, Ce, Fe, Ca)_2(Nb, Ta, Ti)_2 O_6(OH, F, O)$	5.5 Ce	1.14–15.4 Ce 5.88–6.89 Ce	3–4 4.4–4.5
			Cerite	$(Ca, Fe)Ce_3H[(OH)_2/SiO_4/Si_2O_7]$ (? with La, Dy, Al)		20–54 Ce	4.9
14	Silicon	Si	Quartz	SiO_2	46.7 Si		26
22	Titanium	Ti	Rutile	TiO_2	60.0 Ti		4.2
			Ilmenite	$FeTiO_3$	31.6 Ti		4.5

Atom. Nr.	Element		Mineral	Formula	Metal-content in %		Density
					theoretical	actual	
40	Zirconium	Zr	Zircon Baddeleyite	$Zr[SiO_4]$ ZrO_2	49.7 Zr	up to 70 Zr	4.5 4.9–5.4
90	Thorium	Th	Thorite Monazite	$Th[SiO_4]$ see above	71.7 Th	2.02–24.1 Th	4.6 5
23	Vanadium	V	Patronite Descloizite Mottramite Vanadinite Carnotite Roscoelite	VS_4 } $Pb(Zn,Cu)[VO_4/OH]$ $Pb_5[(VO_4)_3/Cl]$ $(K,Na,Ca,Cu,Pb)_2$ $[(UO_2)/(VO_4)]_2 \cdot 3\,H_2O$ V-bearing muscovite	 11.6 V 10.8 V	28–39 V 9.8–13.7 V 11.3–12.8 V 4.38–16.1 V	 6 6.7–7.2 7 2.9–3
41	Niobium	Nb	Columbite Pyrochlor (Koppite) Fergusonite	$(Fe,Mn)(Nb,Ta)_2O_6$ see above $Y(Nb,Ta)O_4$	51.4 Nb	22–54.5 Nb 43–48 Nb 20–32 Nb	5.3–7.3 4.4–4.5 4.3–6.2
73	Tantalum	Ta	Tantalite Fergusonite	$(Fe,Mn)(Ta,Nb)_2O_6$ see above		43–66 Ta 1.6–22 Ta	6.5–8.2 4.3–6.2
24	Chromium	Cr	Chromite	$FeCr_2O_4$	46.4 Cr		4.8
42	Molybdenum	Mo	Molybdenite Wulfenite Powellite	MoS_2 $Pb[MoO_4]$ $Ca[MoO_4]$	60 Mo 26.1 Mo 48 Mo		4.8 7.0 4.4–4.5
74	Tungsten (Wolfram)	W	Ferberite Hübnerite Scheelite	$Fe[WO_4]$ $Mn[WO_4]$ $Ca[WO_4]$	60.5 W 60.7 W 63.8 W	} Wolframite	7.5 7.1 6.0

APPENDIX

Atom. Nr.	Element		Mineral	Formula	Metal-content in %		Density
					theoretical	actual	
92	Uranium	U	Uraninite (Pitchblende)	$(U, Th)O_2$	83.3 U	up to 76.7 U	9.5
88	Radium	Ra	Carnotite	see above		up to 55 U	7
25	Manganese	Mn	Pyrolusite	MnO_2	63.2 Mn		up to 5
			Psilomelane				
			Cryptomelane				
			Polianite				
			Manganite	$Mn_2O_3 \cdot H_2O$	62.46 Mn		4.3
			Braunite	$Mn^{2+}Mn_6^{4+}[O_8/SiO_4]$	63.6 Mn		4.8
			Hausmannite	Mn_3O_4	72.0 Mn		4.7
			Rhodochrosite	$MnCO_3$	47.8 Mn		3.5
			Rhodonite	$Mn[SiO_3]$	41.9 Mn		3.5
			Jacobsite	$MnFe_2O_4$	23.8 Mn		4.7
26	Iron	Fe	Hematite	Fe_2O_3	70.0 Fe		5.2
			Magnetite	Fe_3O_4	72.35 Fe		5.2
			Limonite	$FeOOH$	62.85 Fe		up to 4
			Siderite	$FeCO_3$	48.21 Fe		3.0–3.8
			Chamosite	$(Fe, Fe^{3+})_3[(OH)_2/AlSi_3O_{10}]$ $(Fe, Mg)_3(O, OH)_6$		28.5–37.3 Fe	3–3.4
			Ilmenite	$FeTiO_3$	36.8 Fe		4.5
27	Cobalt	Co	Skutterudite and Smaltite	$CoAs_{3-2}$	28.23 Co		6.5
			Safflorite	$CoAs_2$			6.2
			Cobaltite	$CoAsS$	35.52 Co		6.2
			Linnaeite etc.	$(Co, Ni)_3S_4$		11–53 Co	4.8–5.8
			Asbolane	Co-bearing psilomelane		3.15–27 Co	2–4
			Heterogenite (Stainierite)	$Co(OH)_2$	63.6 Co		2–4

INTRODUCTION TO ORE DEPOSITS

Atom. Nr.	Element		Mineral	Formula	Metal-content in %		Density
					theoretical	actual	
28	Nickel	Ni	Niccolite	NiAs	43.92 Ni		8
			Chloantite and Rammelsbergite	NiAs$_2$	28.14 Ni		6.2–7.2
			Pentlandite	(Fe, Ni)$_9$S$_8$		10–40 Ni	4.5–5
			Garnierite	(Ni, Mg)$_6$[(OH)$_8$ Si$_4$O$_{10}$]		4.3–36.1 Ni	2.3–2.8
	Platinum Metals	Pt	Ferroplatinum	Pt, Fe		75–84 Pt	14–19
44	Ruthenium	Ru				2–4 Ir	
45	Rhodium	Rh					
46	Palladium	Pd	Platiniridium	Pt, Ir		20–67 Pt	20–22
76	Osmium	Os				19–69 Ir	
77	Iridium	Ir	Newjanskite	Ir, Pt, Os, Rh, Ru		1–12 Pt	
78	Platinum	Pt				44–70 Ir	19–21
						17–40 Os	
			Sysserskite	Os, Ir, Ru, Pt, Rh		2–12 Rh	
						7–9 Ru	
						0–12 Pt	
						17–30 Ir	
						38–68 Os	
			Aurosmiridium	Ir, Os, Au, Ru		0–4 Rh	up to 21
						9–14 Ru	
						51 Ir	
						26 Os	
						19 Au 3 Ru	

APPENDIX

Atom. Nr.	Element		Mineral	Formula	Metal-content in %		Density
					theoretical	actual	
78	Platinum	Pt	Osmite	Os, Ir, Rh		80 Os 10 Ir 5 Rh	
			Sperrylite	$PtAs_2$	56.6 Pt		10
			Cooperite	PtS	86.5 Pt		9
			Stibiopalladinite	Pd_3Sb	70 Pd		0.5
29	Copper	Cu	native Copper	Cu	100 Cu		9
			Chalcocite	Cu_2S	79.9 Cu		5–6
			Covellite	CuS	66.5 Cu		4.6
			Chalcopyrite	$CuFeS_2$	34.7 Cu		4.2
			Bornite	Cu_9FeS_6 to Cu_3FeS_3	63.3 Cu to 55.0 Cu		4.9–5.5
			Enargite	Cu_3AsS_4	48.4 Cu		4.4–4.5
			Tetrahedrite	Cu_3SbS_{3-4}		23–45 Cu	
			Tennantite	Cu_3AsS_{3-4}		30–53 Cu	4.4–5.4
			Bournonite	$CuPbSbS_3$	13.0 Cu		5.8
			Cuprite,	Cu_2O	88.8 Cu		6
			Tenorite	CuO	57.5 Cu		4
	Only oxidation zone		Malachite	$Cu_2[(OH)_2/CO_3]$	55.3 Cu		3.8
			Azurite	$Cu_3[(OH)/CO_3]_2$	40.4 Cu		3.3
			Dioptase	$CuSiO_3 \cdot H_2O$	59.0 Cu		3.7
			Atacamite	$Cu_2(OH)_3Cl$	25.5 Cu		2.2
			Chalcantite	$CuSO_4 \cdot 5 H_2O$	56,2 Cu		3.9
			Brochantite	$Cu_4[(OH)_6/SO_4]$			
47	Silver	Ag	native Silver	Ag	100 Ag		10
			Acanthite, Argentite	Ag_2S	87 Ag		7
			Proustite	Ag_3AsS_3	65.4 Ag		5.6

Atom. Nr.	Element	Mineral	Formula	Metal-content in %		Density
				theoretical	actual	
47	Silver[1]	Pyrargyrite	Ag_3SbS_3	59.8 Ag		5.8
		Stephanite	Ag_5SbS_4	68.3 Ag		6.2
		Polybasite	$Ag_{16}Sb_2S_{11}$	75.5 Ag		6
		Pearceite	$Ag_{16}As_2S_{11}$	78.4 Ag		6.1
	Only oxydation zone	Cerargyrite	$AgCl$	75.0 Ag		5.5–5.6
79	Gold[2] Au	native Gold	Au, Ag		80–98 Au	15–19
		Electrum	Ag, Au		70–75 Au	12–16
		Calaverite	$AuTe_2$	43.7 Au		9
		Sylvanite	$AuAgTe_4$	24.2 Au		8
		Nagyagite	$AuTe_2 \cdot 6\,Pb(S, Te)$		6–13 Au	6.8–7.5
		Petzite	$(Ag, Au)_2Te$		up to 75 Au	8.7–9
30	Zinc Zn	Sphalerite	$(Zn, Fe, Mn, Cd)S$ pure ZnS		up to 67 Zn	3.9–4.1
		Wurtzite			67.1 Zn	
		Franklinite	$(Zn, Mn)(Fe, Mn)_2O_4$		7–20.5 Zn	5
		Willemite	Zn_2SiO_4	58.6 Zn		4
		Zincite	ZnO	80.3 Zn		5.5
	Only oxidation zone	Smithsonite	$ZnCO_3$	52.1 Zn		4.3
		Hemimorphite	$H_2Zn_2SiO_5$	54.3 Zn		3.5
48	Cadmium Cd	Sphalerite	$(Zn, Fe, Mn, Cd)S$		0.05–3.2 Cd	3.9–4.1
		Smithsonite	$(Zn, Cd)CO_3$		0.02–0.8 Cd	4.3–4.4
80	Mercury Hg	Cinnabar	HgS	86.2 Hg		8
		Schwazite (Hg-tetrahedrite)	$(Cu_2, Hg)_3Sb_2S_6$		up to 17 Hg	5

[1] The main silver production is from: galena, tetrahedrite chalcocite, pyrite whose Ag-concentration varies widely.
[2] Other main gold-ores are pyrite, arsenopyrite, stibnite etc.

APPENDIX

Atom. Nr.	Element		Mineral	Formula	Metal-content in %		Density
					theoretical	actual	
31	Gallium	Ga	Germanite	$Cu_3(Fe, Ge, Ga)S_4$		up to 0,8 Ga	4.3
			Gallite	$CuGaS_2$		?	4.4
			Sphalerite	see above		traces	4.0
49	Indium	In	Sphalerite	see above		traces	
81	Thallium	Tl	Pyrite	see below		traces	
32	Germanium	Ge	Germanite	$Cu_3(Fe, Ge, Ga)S_4$		up to 8 Ge	4.3
			Renierite	$Cu_3(Fe, Ge)S_4$, Fe-rich Germ.			± 4.3
50	Tin	Sn	Cassiterite	SnO_2	78.7 Sn		6.8
			Stannine	Cu_2FeSnS_4	27.6 Sn		4.4
			Cylindrite	$Pb_3Sn_4Sb_2S_{14}$ (?)	24.8 Sn		5.4
			Teallite	$PbSnS_2$	30 Sn		6.4
82	Lead	Pb	Galena	PbS	86.6 Pb		7.5
			Bournonite	$CuPbSbS_3$	42.4 Pb		5.8
			Boulangerite	$Pb_3Sb_4S_{11}$	55.2 Pb		6.1
	Only oxi-	{	Cerussite	$PbCO_3$	77.55 Pb		6.5
	dation zone		Anglesite	$Pb[SO_4]$	68.33 Pb		6.3
			Pyromorphite	$Pb[Cl/(PO_4)_3]$	76.38 Pb		7.0
33	Arsenic	As	Arsenopyrite	$FeAsS$	46.0 As		6
			Loellingite	$FeAs_2$	72.8 As		7
51	Antimony	Sb	Stibnite	Sb_2S_3	71.7 Sb	75 Sb	4.5
	oxidation zone		Antimonyochre	$(Ca, NaH)Sb_2O_6(O, OH, F)$			5.1–5.2
83	Bismuth	Bi	native Bismuth	Bi	100 Bi		9.8
			Bismuthinite	Bi_2S_3	81.3 Bi		6.5

Atom. Nr.	Element		Mineral	Formula	Metal-content in %		Density
					theoretical	actual	
16	Sulfur	S	native Sulfur	S	100 S		2
			Pyrite	FeS_2	53.4 S		5.2
			Pyrrhotite	$Fe_{1-x}S$	36.5 S		4.6
			Gypsum	$Ca[SO_4] \cdot 2H_2O$	23.2 S		2.3
34	Selenium	Se	Pyrite	see above		traces	
			Sphalerite	see above		traces	
			Selenides				
52	Tellurium	Te	Calaverite	$(Au, Ag)Te_2$	56.4 Te		9.0–9.4
			Sylvanite	$AuAgTe_4$	62.6 Te		7.9–8.3
			Nagyagite	$AuTe_2 \cdot 6Pb(S, Te)$ (?)		18–30 Te	6.8–7.5
			Petzite	$(Ag, Au)_2Te$	24.4 Te		8.7–9.0
			Hessite	Ag_2Te	37.1 Te		8.2–8.9

Index

abundance ratio (Clarke number), 3
Algeria, North Africa, 40
allochemical (metasomatic) reactions in metamorphism, 105
Altenberg, East Germany, 26, 36
aluminium, deposits of, 81, 82
antimony-mercury formation, 32, 62
assimilation, 10, 15
atmosphere, importance of, in exogenetic cycle, 63-64

bacteria, importance of, in formation of ore deposits, 75, 87-88, 96-99
Banat, Rumania, 27-28, 40
Barnes, H. L., 31
Barsukov, V. L., 23
Barton, P., 29
basalt, alteration of, 82-83
  differentiation of, 6-8
basic igneous rock, weathering of, 83
Bateman, A. M., 43
bauxite, formation of, 73, 81-82
  metamorphism of, 107, 111
Bentz, A., 77, 78
Bilbao, Spain, 40
Bingham, Utah, USA, 42, 53
biosphere, importance of in exogenetic cycle, 63-64, 70, 74-75
Birmingham, Alabama, USA, 95
bismuth-cobalt-nickel-silver formation, 32, 37, 61
Black Sea, 98
bog iron ore, formation of, 88-91
Borchert, H., 7, 58-59, 91-92
Bosnia, Yugoslavia, 40
Bowen, reaction series of, 9
Brittany, France, 95
Broken Hill, Australia, 112-113
Broken Hill, Zambia, 81
Bushveld Massif, South Africa, 15-19
Butte, USA, 55

Campagna, 98
carbonaceous rock, metamorphism of, 111
carbonate deposit, 74
carbonate formation, 36-37
  metasomatic replacement of, 38-42
cassiterite, formation of, 23-25
  in detrital sediments, 71-76
cementation zone in weathered ore deposits, 80-81

chalcopyrite, formation of in basic igneous rocks, 15, 60
Chalilovo, USSR, 85
chamosite, metamorphism of, 111
  -thuringite, formation of, 93-95
chromite, deposits of, 15-17
Chuquicamata, Chile, 43
Clarke number (abundance ratio), 3
classification of ore deposits, 4-5
  of sedimentary deposits, 69-71
clay ironstone, formation of, 92-93
Clifton Morenci, USA, 52
climatic influence on soil profile, 67
  influence on weathering, 66-68
Climax, USA, 52
Cloos, H., 54
Cobalt City, USA, 61
Cœur d'Alene, USA, 55, 81
colloidal solution in hydrothermal systems, 29-30
  in sedimentary processes, 73-74, 93
Colorado Plateau fault zone, 53
complex solution in hydrothermal systems, 29
Conakry, Guinea, 83
copper, deposits of, 42, 45-47, 50, 88
  -iron-arsenic formation, 32
Cornwall, England, 25, 50
corundum, formation of, 107-111
Cotta, B., 27
crust, division of, 6
crystallization differentiation, 8-9, 15
  fractional, 60
  products of, 11-12
  sequence of, 9, 61
Cumberland, England, 42

Daubre, 23
Deccan Plateau, India, 82
deflocculation of colloidal solutions, 73-74
deposition in sedimentary processes, 68-70
deposits, see Ore deposits
diamond placers, 76-78
differentiation, magmatic, 8-9, 13-19, 57
discontinuities within the crust, 6-8
dolomite, metasomatic replacement of, 24
Durham, England, 103
Dzheskasgan, USSR, 88-89

electrochemical series, 81
element content of hydrothermal solutions, 29-31

element enrichment, within detrital sedimentary deposits, 71, 87
  within kupferschiefer, 101
  within limonitic oolitic deposits, 93
  within nickel silicate deposits, 83
  within organically precipitated deposits, 96-98
  within oxidate and hydrolysate minerals, 73
  within pisolitic limonites, 83
  within pneumatolytic ore deposits, 21, 23
endogenetic cycle, 5-6, 42
equilibrium relation between iron oxides and sulphides, 31
Erzgebirge, East Germany, 24, 26, 33-37, 61, 78
evaporite deposit, 74
exogenetic cycle, 5, 42, 63-71

fabric, alteration of kinetic stress, 107
Falun, Sweden, 61, 108
feldspar, decomposition of, 81
Flin-Flon, Manitoba, Canada, 111
flocculation of hydroxides, 83, 92
fluorite-barite formation, 36-37, 56, 61
  formation by pneumatolysis, 23-25
fractionation of elements in the exogenetic cycle, 64-66
Franken, West Germany, 95
Franklin, New Jersey, USA, 113
Freiburg, West Germany, 33-38, 55
Freihung, Oberpfalz, DBR, 100

gabbro, weathering of, 83
Gaellivare, N. Sweden, 45
galena, deposits of, 44, 101, 109
gangue minerals, 31
garnet in detrital sediments, 71, 73, 76
garnierite, deposits of, 83-85
gel structures within hydrothermal deposits, 30
  sedimentary transportation of, 73-74, 93
Geyer, East Germany, 26
Ghana, 108
goethite, deposits of, 40, 42
gold-silver formation, 32, 62
  within detrital deposits, 73, 76, 79
Goranson, R. W., 10
gossan, formation of, 79
Graengesborg, Sweden, 45
granitic magma, origin of, 6-8
greenalite, deposits of, 108
greisenization, 23-27
gypsum, formation and deposits of, 74-75, 98

Hagendorf, W. Germany, 22
Halimba, Hungary, 82

halmyrolysis, *see* weathering
hausmannite, formation of, 111
hematite, pneumatolylic formation of, 23-25
  stability of, 31
Hessen, West Germany, 102
Hohe Tauern district, Alps, 52
Huettenberg, Carinthia, Austria, 40, 56
Hunsrueck Mountains, W. Germany, 83
hydrogen sulphide, formation by organisms, 96-98
hydrosphere, importance of in exogenetic cycle, 63-64, 70
hydrothermal deposit, recrystallization of by static-kinetic metamorphism, 107
  deposit, structural form of, 32
  solution, composition of, 29-31
  stage, 12-13

ilmenite in detrital sediments, 71, 73, 76
impregnation, definition of, 4
  deposits, 42
insolation, 65
intramagmatic accumulation, 51
ionic potential of elements, 65
  solution in hydrothermal systems, 29-30
iron minerals, equilibrium relations of, 31
  -manganese-barium formation, 32, 37
  -nickel formation, 32
  -ore deposits, 44, 45, 48, 83, 93-95, 108-109
isochemical reactions in metamorphism, 105
itabiritic iron-ore deposits, 108

Johannesburg, South Africa, 15
Jutland, Denmark, 91

kaolinite, formation of, 73, 81
Katanga, Congo, 81, 99, 100
Keban, Turkey, 55
Kiruna, N. Sweden, 45, 48
Kraume, E., 110
Krauskopf, K. B., 29
Kremze, Czechoslovakia, 85
Krivoi Rog, Ukraine, USSR, 108, 110
Kullerud, G., 31
Kupferschiefer, 98-102

Lahn Dill type iron-ore deposits, 44, 52, 60-61, 108
Laisvall, N. Sweden, 100
lake ore, formation of, 80-91
Lake Superior district, N. America, 95, 108-109
laterite, formation of, 73, 81-83
Laurion, Greece, 55

INDEX

lead deposits of, 44-47, 55, 101, 109
  -zinc-silver formation, 32, 36, 62
Leadville, USA, 52-53
Leksdal, Norway, 61, 111
Les Beaux, France, 82
limestone, formation of, 74-75
  pneumatolytic replacement of, 24-25, 28
limonite, metamorphism of, 107, 111
  pisolitic and oolitic, 83, 93-95
liquid immiscibility of basic magmas, 9, 13
liquid magmatic stage of crystallisation, 12-13
lithosphere, importance of in exogenetic cycle, 63-64, 70
Lökken, Sweden, 111
Lorraine-Luxemburg area, 95
Lower Silesian-Bohemian basin, 87

MacDiarmid, R. A., 50, 86, 112
magma, origin of, 6-8
magmatic differentiation, 8-19
magmatism, types of, 8, 44, 57-62
magnesite, formation by weathering, 85
  metasomatic origin of, 106
magnetite, deposits of, 45, 61, 108-109
  formation of, 23-28
  in detrital sediments, 71, 76, 78
  stability of, 31
Mansfeld, East Germany, 102-103
Martini, H. J., 77
Maubach-Machernich area, Eifel, W. Germany, 99-100
Mayari, Cuba, 83
Meggen, W. Germany, 45, 47
Meinkjar, Norway, 18-19
Merensky, H., 17
Mertie, J. B., Jr., 20
metal based classification of ore deposits, 4
metamorphism, transformation of deposits by static kinetic, 107-111
  transformation of deposits by contact, 111
  metasomatic replacement, process of, 38-42
Miami, Colorado Plateau, USA, 52
Minas Geraes, Brazil, 108
mineralization, cycles of, 35-38
minerals, pegmatitic/pneumatolytic, 21
  within detrital sediments, 71, 76
monazite, placer deposits of, 76-78
Morocco, N. Africa, 40

Nagpur, India, 108
New Caledonia, 85-86
nickel, deposits of, 19-20, 83-86
Nicopol, Ukraine, USSR, 96-97
Niggli diagram, 12
Niggli, P., 11-12, 72

Nishni Tagil, Urals, USSR, 76
Normandy, France, 95

Oberpfalz, W. Germany, 22
Oelsner, O., 55
olivine, weathering of, 83
oolite, formation of, 93
ore deposit, Calabria type, 98-99
  definition and classification of, 1-5
  detrital, 70, 75-76
  disseminated, 4, 15
  formation of in endogenetic cycle, 6-62
  formation of in exogenetic cycle, 75-103
  hydrothermal formation of, 28-48, 61-62
  impregnation, 42, 61
  limnic, 69-70, 73
  liquid magmatic (intramagmatic) formation of, 13-19, 51
  marine, 69, 73
  marine precipitated, 100-103
  metamorphic, classification of, 107
  metamorphic, transformation of, 104-113
  metasomatic replacement, 38-42, 51, 61
  metasomatic siderite, 39-40
  of sialic magmatism, 61-62
  of simatic magmatism, 60-61
  organically precipitated, 96
  pegmatitic/pneumatolytic formation of, 20-28, 61-62
  placer, 70, 73, 75-79
  precipitated sedimentary, 86-103
  red bed, 86-88
  salt, 70-74
  sedimentary iron and manganese, 91-96
  submarine hydrothermal, 42-44, 51-52
  terrestrial, 69-73
  topomineral reaction, 51
  variation according to level of intrusion, 50-51, 60-61
  weathering of, 79-81
Ore mountain, Erzberg, Austria, 39
orogenesis, relation to ore deposit formation, 57-62
Orsk, Urals, USSR, 85
Outukumpu, Finland, 111
oxidation, zone of, 73, 79-80

palingenesis, 57-62
paragenesis, of ore deposits, 30-32, 51
Park, Ch. F., Jr., 50, 86, 112
Pechtelgrun, Vogtland, 24
pegmatitic-pneumatolytic phase of magmatism, 12, 20-28
Peine-Ilsede, W. Germany, 76
peridotite, lateritic weathering of, 83-86
Persian Gulf, 74
Petrascheck, W. E., 53
petroleum deposits, 70-71, 73

pH, importance of in sedimentary processes, 63, 65, 69
platinum, magmatic deposits of, 19-20
  placer deposits of, 76-77
Postmasburg, S. Africa, 108
Prague trough, Czechoslovakia, 95
precipitation, gravitative, in simatic magmas, 13-15, 60
pyrite, deposits of, 42, 44-47, 101, 109
  equilibrium relations of, 31
  metamorphism of, 111
  sequence in ore deposition, 36
pyrrhotite, equilibrium relation of, 31
  nickeliferous, deposits of, 19

Rammelsberg, DBR, 108-110
Rectoren deposit, 45, 48
redox potential, importance of sedimentary processes, 63, 69, 92-93, 101
Richelsdorf, DBR, 102
Rio Tinto–Huelva district, Spain, 44, 51, 60-61
Rio Tinto–Meggen type sulphide deposits, 44, 52, 60-61
Roros, Norway, 61, 111
Rote Fäule, Kupferschiefer, 101-102
Rougu-Norauda, Quebec, Canada, 111
Rumania, 27

Sadisdorf, W. Germany, 26
St. Egidien, W. Germany, 85
Salzgitter, W. Germany, 76
sapropel facies, 70-73, 98-99
Saugerhausen trough, W. Germany, 102
scheelite, pneumatolytic formation of, 24-25
schlieren, 4, 9, 15, 21
Schmalkalden, East Germany, 39-40
Schneiderhöhn, H., 12, 16-17, 22, 28, 40-41, 48
sediments, classification of, 69-75
segregation, magmatic, 12-15
serpentinite, weathering of, 83-85
Shearman, D. J., 74
siderite, metasomatic deposits of, 39-42
siliceous rock, formation of, 73-74
silver-rich sequence in hydrothermal ore formation, 36-37
Singhbhum district, India, 109
Skacel, J., 44
skarn, 24, 45
Smirnov, V. I., 66, 89, 110
Smyrna, Peru, 111
soil, limonitic, 88-91
  profiles, showing dependency on climate, 67
solubility, of H_2O in albite melts, 10
  of sulphides, 29, 101
  series of, 65-66

sphalerite, deposits of, 44-47, 101, 109
stabilities, fields for iron minerals, 31
Sternberk-Hor area, Benesov, CSSR, 27
Stille, H., 8, 57
stockscheider, 21, 27, 52
Styrian ore mountain, Austria, 41
Sudbury, Ontario, Canada, 19-20, 85
Sulitjelma, Sweden, 111
sulphide-carbonate sequence in hydrothermal ore formation, 36
  melts, 9, 15
  oxidation of, 79-81
sulphur, cycle, 98-99
  native, formation of, 79-81

tectonic control of ore deposits, 52-56
telescoping, of mineral paragenesis, 49, 51, 60
temperature-concentration diagram, 12
  -pressure diagram, 12
Thuringia, East Germany, 95
tin, deposits of, 25-27, 50, 78
  deposits, pneumatolytic formation of, 22-25
  -tungsten formation in hydrothermal ore deposits, 35-36, 62
topaz, pegmatic/pneumatolytic formation of, 21-25
tourmaline, pegmatitic/pneumatolytic formation of, 21-25
transportation in sedimentary processes, 68-70
Trieben, Austria, 55
Tröger, 7
Tshiaturi, Caucasus, USSR, 96
Tsumeb, S.W. Africa, 81
Tunisia, N. Africa, 40

ultrabasic rock, weathering of, 83-85
uranium, deposits of, 79, 90
  -iron formation in hydrothermal ore deposits, 36, 62
  metallization in red bed deposits, 88
  -quartz sequence in hydrothermal ore deposits, 36

vanadium, deposits of, 88, 111
Vaskö, Banat, 28
Verhoogen, J., 29
Vogt, J. H. L., 18-19

water content of magma, 9-11
weathering, formation of ore deposits by, 75-86
  types of, 64-68
White Pine, Michigan, 100
Williams, D., 46

Witwatersrand, S. Africa, 78-79
wolframite pegmatitic/pneumatolytic formation of, 24-27
workability of ore deposit, 2-3
Wuerttemberg, W. Germany, 83

Zajaczek—Grodzica, Poland, 103
Zeschke, G., 109

zinc, deposits of, 44-47, 55, 101, 109
 -tin-copper sequence within hydrothermal ore deposits, 36
zircon, enrichment in detrital sediments, 71
zonal distribution of earth's crust, 6-8, 38, 48-50
zones of formation of sedimentary iron and manganese deposits, 92, 96-97